Springer Tracts in Mechanical Engineering

Springer Tracts in Mechanical Engineering (STME) publishes the latest developments in Mechanical Engineering - quickly, informally and with high quality. The intent is to cover all the main branches of mechanical engineering, both theoretical and applied, including:

- Engineering Design
- Machinery and Machine Elements
- Mechanical structures and Stress Analysis
- Automotive Engineering
- Engine Technology
- Aerospace Technology and Astronautics
- Nanotechnology and Microengineering
- Control, Robotics, Mechatronics
- MEMS
- Theoretical and Applied Mechanics
- Dynamical Systems, Control
- Fluids mechanics
- Engineering Thermodynamics, Heat and Mass Transfer
- Manufacturing
- Precision engineering, Instrumentation, Measurement
- Materials Engineering
- Tribology and surface technology

Within the scopes of the series are monographs, professional books or graduate textbooks, edited volumes as well as outstanding PhD theses and books purposely devoted to support education in mechanical engineering at graduate and post-graduate levels.

Indexed by SCOPUS and Springerlink. The books of the series are submitted for indexing to Web of Science.

To submit a proposal or request further information, please contact: Dr. Leontina Di Cecco Leontina.dicecco@springer.com or Li Shen Li.shen@springer.com.

Please check our Lecture Notes in Mechanical Engineering at http://www.springer.com/series/11236 if you are interested in conference proceedings. To submit a proposal, please contact Leontina.dicecco@springer.com and Li.shen@springer.com.

More information about this series at http://www.springer.com/series/11693

Bruno Chanetz · Jean Délery · Patrick Gilliéron ·
Patrick Gnemmi · Erwin R. Gowree ·
Philippe Perrier

Experimental Aerodynamics

An Introductory Guide

 Springer

Bruno Chanetz
ONERA
Meudon, France

Jean Délery
ONERA
Meudon, France

Patrick Gilliéron
French Aeronautical and
Astronautical Society
Paris, France

Patrick Gnemmi
ISL
Saint-Louis, France

Philippe Perrier
Dassault Aviation
Paris, France

Erwin R. Gowree
ISAE-SUPAERO
Toulouse, France

ISSN 2195-9862   ISSN 2195-9870   (electronic)
Springer Tracts in Mechanical Engineering
ISBN 978-3-030-35564-7   ISBN 978-3-030-35562-3   (eBook)
https://doi.org/10.1007/978-3-030-35562-3

This Springer imprint is published by the registered company Springer Nature Switzerland AG
The registered company address is: Gewerbestrasse 11, 6330 Cham, Switzerland

Foreword

I had a great pleasure reading this book about experimental aerodynamics.

It stems largely from the skills gathered within the Aerodynamics Technical Committee of the French Aeronautical and Astronautical Society (3AF); skills valued by what is called "Collective Intelligence".

All the contributions were organised by Jean Délery, who directs with passion and with a keen interest for pedagogy this Technical Committee, whose qualities are recognised worldwide.

For all those who express an interest or a curiosity for the subject, I recommend reading this book.

They will discover a broad and diverse range of fields of application of aerodynamics, as well as the history and the state of the art of the experimental devices and their ever-increasing capabilities. Readers will also understand that the oppositions that fuelled the discussions a few years ago, between experimental studies and numerical simulations, no longer have any reason to exist: there is perfect complementarity, even enrichment, for the best representation of physical phenomena, while enriching our knowledge.

Engineers in charge of developing new products will find in this book the information necessary for the success of their work.

I wish that many people will appreciate this book, and that they all take as much pleasure and interest as when I read it.

This one must find the best spot in the library of an engineer.

Michel Scheller
President of the French Aeronautical
and Astronautical Society

Preface

The purpose of this book is to present an inventory of experimental facilities and techniques commonly utilised in the field of aerodynamics, which remains paramount for the design of aerial and ground vehicles, propulsion and energy generation systems, and, also in the civil engineering industry and in more fundamental environmental flow studies, but not limited to these. Aerodynamic studies do not only target the improvement of vehicle performance and comfort, but also, more and more, the emission of greenhouse gases within the earth's atmosphere and noise within the neighbourhood of residential areas. Due to the ever-increasing demand of both aerial and ground transportation, these emissions require immediate moderation in order to sustain this industry, and improvement in aerodynamic performance is definitely a way forward. The book also aims to provide updated information on the means and techniques used by aerodynamicists, and more generally in the field of fluid mechanics. In particular, a large part is dedicated to measurement techniques and instrumentations which have undergone tremendous improvements over the last 40 years. These improvements Allowing access to very precise diagnostic techniques and hence increasing our knowledge of the behaviour of highly complex flows. This remarkable development can be compared with the equally impressive progress in the field of computational fluid dynamics (CFD). Measurement and instrumentation is a stimulating field in its own right and has greatly benefited from the most advanced knowledge of basic physics, mainly wave theory, optics and signal processing, and mathematics. This book, however, retains a general character; the readers wishing to deepen their knowledge in the subject are referred to more specialised literature in the field of measurement techniques. The targeted readers are graduate students and engineers wishing to embark on the theme of experimental aerodynamics.

After a presentation of the objectives and a brief introduction to methods of experimental aerodynamics, as well as related issues and limitations in Chaps. 1 and 2, the book devotes a large part to the description of wind tunnels and measurement techniques. It is not a catalogue of existing facilities but a presentation of some of the most typical wind tunnels in specific fields of application (aeronautics, space exploration, automobile, railway, energy production, civil engineering, etc.) predominantly in France, but also in other countries. However, there is no absolute demarcation between facilities, as "aeronautical" wind tunnel can be also used to test ground vehicles or wind turbines, for example. Emphasis is placed on the particular problems encountered during design, production and operation, accordingly with the simulated speeds, ranging from low subsonic to hypersonic. It is appropriate to distinguish between industrial wind tunnels intended for the development of prototypes, by testing either scaled-down models or full-scale vehicle itself, and the research wind tunnels devoted to the detailed study of particular phenomena such as separation, laminar-to-turbulent transition, etc. However, there may be an overlap between the two types of facilities.

Chapter 3 deals with subsonic wind tunnels, while presenting a wide range of facilities to cover the needs of aeronautics, automotive and civil engineering industries, energy production, etc. We are also interested in presenting facilities dedicated to study the effects of adverse weather conditions such as rain, ice and snow on ground vehicles or ice accretion on aircraft. Also an extensive survey of aeroacoustic wind tunnels whose purpose is to characterise the noise generated by the flow over the overall vehicle or individual parts such as aircraft landing gears, flap and control surfaces, jet engines, side mirrors or other isolated surfaces.

Transonic wind tunnels in Chap. 4 occupy a strategic place as they are primordial for commercial and business aviation, as well as military aircraft both manned and unmanned combat aerial vehicle (UCAV). The transonic regime is also of interest for turbomachinery and jet engines, and high-speed trains' tunnel entry. These facilities are rather limited because of particular design and operation challenges related to the occurrence of complex flow phenomena while operating in the vicinity of the speed of sound.

Chapter 5 deals with supersonic wind tunnels which cover the needs of high-speed applications for the design and optimisation of fighter aircraft, missiles and ammunition and space launchers while flying within the earth's atmosphere. Other applications where supersonic regime is encountered are engine intakes while operating at maximum power, supersonic intakes and nozzles, and other instances when the flow accelerates rapidly due to large convergences and curvatures.

Hypersonic flows covered in Chap. 6 are of great interest again in military applications such as for the design of hypervelocity tactical missiles, strategic missiles and projectiles. In the space launchers and exploration field, this regime is encountered during the re-entry of vehicles into earth's atmosphere but also in other atmospheres such as those of Mars and Venus. Space launchers are also of concern while exiting Earth's atmosphere at high altitude. The realisation of hypersonic testing requires the generation of flows not only at high Mach number but also with high specific energy. Because of the very high speeds encountered during the atmospheric re-entry, high levels of thermodynamic effects are present.

Much of the book is devoted to the means of diagnosis and characterisation of flows, including very valuable visualisation techniques presented in Chap. 7, and the measurement of the aerodynamic forces and moments exerted on the vehicle by the fluid in motion is addressed in Chap. 8. The determination of the surface properties such as pressure, skin friction and heat transfer is of fundamental importance in most applications; the techniques are visited in Chap. 9. In addition to the traditional measurement by static pressure at the wall using pressure tappings, the use of more recently developed Pressure Sensitive Paints (PSP) allows for the global wall pressure distribution over a larger and sometimes whole part of the model to be captured without the need for discrete intrusive, expensive and cumbersome pressure sensors and data acquisition system. Similar progress is made in the measurement of heat transfer while increasingly competing with optical methods like infrared thermography or the use of temperature-sensitive paints (TSP).

The local properties of the flow (pressure and velocity in particular) are still determined by standard means such as pressure probes or thermocouples for the stagnation pressure and temperature. Hot wire is still widely used for time-averaged velocity measurements, and due to its very short response time, it is still the most reliable technique for measurement of velocity fluctuations and turbulence scales, and this will be covered in Chap. 10.

Chapter 11 presents the so-called non-intrusive techniques for in situ measurement of flow properties, mainly velocity. This is an essential part of the book. These means of characterisation of flows have undergone a real breakthrough with the advent of more affordable lasers and remarkable advances in the field of optronics and data processing. At first, laser Doppler velocimetry (LDV) and then particle image velocimetry (PIV) have been essential tools to capture and resolve very complex flow phenomena and have now become commonplace in almost all aerodynamics laboratories.

The development of techniques based on the excitation of molecules or atoms of the gas by laser or electron beam has allowed the development of methods without the need to seed the flow by particles, unlike in LDV and PIV. Spectroscopic techniques give access to the gas properties such as pressure, temperature, density and also to its velocity and composition in the case of reactive flows. They are widely used in the study of hypersonic flows and will be discussed in Chap. 12.

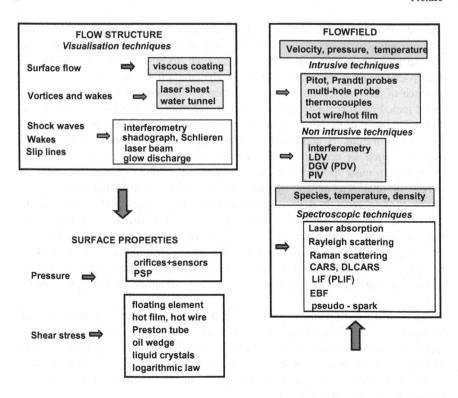

The diagram above summarises the methodology to be used while analysing a flow, from the visualisations giving an idea of its global structure, through global measurements of forces, and then wall properties moving towards detailed exploration of the local quantities. The methods listed in the table are detailed as the main content of the book. Chapter 13 is devoted to ongoing developments to build a methodology that tightly couples experiments and numerical simulations. The purpose of this innovative approach is to take into account the spurious effects from wind tunnel testing, to supplement measurements and to ensure close coupling to optimise model design, test preparation, wind tunnel operation and interpretation of results. This is the computer-assisted wind tunnel, associated with the reconstruction, or assimilation, of data whose goal is to reconstitute, with the help of CFD, a field from dispersed measurements. This area has known many important developments.

Finally, Chap. 14 proposes a review of the status quo of experimental aerodynamics by trying to project a vision of this discipline and to define the problems it will face in the years to come.

On such a vast topic, it is difficult to establish an exhaustive bibliography. For this reason, only general publications on experimental aerodynamics and publications highlighting recent and outstanding work in the facilities described have been selected. For the measurement techniques part, the large number of publications will allow the reader to refer to a more complete and detailed technical documentation.

Meudon, France	Bruno Chanetz
Meudon, France	Jean Délery
Paris, France	Patrick Gilliéron
Saint-Louis, France	Patrick Gnemmi
Toulouse, France	Erwin R. Gowree
Paris, France	Philippe Perrier

Acknowledgements

Jean-Charles Abart, ONERA
Michel Alaphilippe, ONERA
Sandrine Aubrun, PRISME Laboratory
Pascal Audo, ONERA
Bruno Bassery, IAT
Enrico Bergamini, IAT
Patrick Berterretche, PPRIME Institute
Benoit Blanchard, CSTB
Henri Boisson, IMFT
Jean-Paul Bonnet, PPRIME Institute
Jacques Borée, PPRIME Institute
Jean-Paul Bouchet, CSTB
Guy Boyet, ONERA
Marianna Braza, IMFT
Jean-François Bret, ONERA
Jean-Marc Breux, PPRIME Institute
Eric Brunel, ONERA
Reynald Bur, ONERA
Yves Carpels, ONERA
Daniel Caruana, ISAE-SUPAÉRO
Clotilde Chagny, IAT
Patrick Chassaing, IMFT
Marie-Claire Coët, ONERA
Jean Collinet, ArianeGroup
Christophe Corre, LMFA
Federico De Filippis, CIRA
Christophe Deltreil, IAT
Jean-Michel Desse, ONERA
Philippe Devinant, PRISME Laboratory
Pierre Dupont, IUSTI

Jean-Paul Dussauge, IUSTI
Biagio Esposito, CIRA
Denis Gély, ONERA
Xavier Gergaud, ACE
Yves Gervais, PPRIME Institute
Steve Girard, PPRIME Institute
Francesco Grasso, IAT
Gilles Harran, IMFT
Vincent Herbert, PSA-Peugeot Citroën
Émilie Jérôme, DGA Aero-engine Testing
Peter Jordan, PPRIME Institute
Laurent Keirsbulck, Polytechnic University of Hauts-de-France
Viviana Lago, ICARE Institute
François Lambert, ONERA
Janick Laumonier, PPRIME Institute
Anton Lebedev, PPRIME Institute
Benjamin Leclaire, ONERA
Friedrich Leopold, ISL
Yves Le Sant, ONERA
Jan Martinez Schramm, DLR
Marie-Claire Mérienne, ONERA
Bruno Mialon, ONERA
Francis Micheli, ONERA
Pascal Molton, ONERA
Christophe Noger, IAT
François Paillé, PPRIME Institute
Sébastien Piponniau, IUSTI
Thierry Pot, ONERA
Jacques Pruvost, ONERA
Antonio Ragni, CIRA
Philippe Reijasse, ONERA
Piergiovanni Renzoni, CIRA
Steve Schneider, Purdue University
Christophe Sicot, PPRIME Institute
Denis Sipp, ONERA
Jean Tensi, 3AF
Jean-Pierre Tobeli, ONERA
Eduardo Trifoni, CIRA
Christophe Verbeke, ONERA
Rémi Vigneron, GIE S2A

Contents

About the Authors

Bruno Chanetz has been a research engineer at ONERA since 1983. He holds a Ph.D. from Lyon I University and is qualified to director of research (HDR). He is the author of numerous publications on hypersonics and plasmas for aerodynamics. He was responsible for the wind tunnels at ONERA Meudon. Besides he was in charge of the boundary layer course at the Ecole Centrale de Paris in 1996, and he was appointed as Associate Professor at the University of Versailles Saint-Quentin-en-Yvelines in 2003, then at the University Paris-Ouest in 2009, where he still teaches.

Jean Délery is an engineering graduate from SUPAERO who joined ONERA in 1964 where he participated in major French and European aerospace programmes of the time. Specialist in the field of experimental aerodynamics, he is the author of numerous scientific publications and several books. He held teaching positions in particular at the University of Versailles Saint-Quentin-en-Yvelines, Sapienza University of Rome, SUPAERO and Ecole Polytechnique and is qualified to director of research (HDR). He is currently Emeritus advisor at ONERA and Chairman of the Aerodynamics Technical Committee of 3AF.

Patrick Gilliéron is an aerodynamic engineer, with a Ph.D. in mechanics and a director of research (HDR). Author of several international publications, he has worked for more than 35 years on the analysis, understanding and control of flow separation. He has been involved as a teaching fellow and then associate professor at the National Conservatory of Arts and Crafts (CNAM) from 1987 to 2001. He created and directed a research group in "Fluid Mechanics and Aerodynamics" from 2002 to 2011 at the Research Department of Renault.

Patrick Gnemmi is an engineer, holds a Ph.D. in mechanics and is qualified to director of research (HDR). For more than 35 years, at the French-German Research Institute of Saint-Louis (ISL), he has carried out research in aeroacoustics of helicopter rotors, aerodynamics and aerothermodynamics of projectiles and missiles including their means of guidance. At present, he is responsible for ISL's Aerodynamics and Outdoor Ballistics group. He is the author of dozens of international publications.

Erwin R. Gowree is an associate professor in applied aerodynamics at ISAE-SUPAERO. He holds a Ph.D. in aerodynamics from City, University of London, where he stayed on as a Research Fellow for few years before joining Rolls Royce as an aeroelasticity specialist. His research activities are mainly in experimental aerodynamics, focussing on stability and transition of flows while also diversifying into bio-inspired aerodynamics. He is in charge of the applied aerodynamics and the experimental practice module at ISAE-SUPAERO.

Philippe Perrier is a graduate of the Ecole Centrale de Paris and has spent more than 40 years at the Dassault Aviation Design Office. He has been involved in the aerodynamic design of all of the company's aircraft since the 1970s, from the Falcon 50 to the Falcon 5X, the Rafale and the nEUROn. He was Technical Director of the Rafale Program and Technical Vice President in charge, in particular of aerodynamic design. Philippe Perrier holds the Medal of Aeronautics.

Acronyms

3AF	Association Aéronautique et Astronautique de France (French Aeronautics and Astronautics Society)
ACE	Aero Concept Engineering
ASI	Agenzia Spaziale Italiana
ASL	Airbus Safran Launchers
BETI	Bruit-Environnement-Transport-Ingénierie
CARS	Coherent Anti-Stokes Raman Scattering
CCA	Constant Current Anemometry
CEA	Commissariat à l'énergie atomique et aux énergies alternatives (French Alternative Energies and Atomic Energy Commission)
CEAT	Centre d'Etudes Aérodynamiques et Thermiques
CFD	Computational Fluid Dynamics
CIRA	Centro Italiano Ricerche Aerospaziali
CNAM	Conservatoire National des Arts et Métiers
CNRS	Centre National de la Recherche Scientifique (French National Centre for Scientific Research)
CROR	Counter-Rotating Open Rotor
CSTB	Centre Scientifique et Technique du Bâtiment (Scientific and Technical Centre for Building)
CTA	Constant-Temperature Anemometry
CVA	Constant-Voltage Anemometry
DARPA	Defence Advanced Research Projects Agency
DES	Detached Eddy Simulation
DGV	Doppler Global Velocimetry
DLCARS	Double-Line Coherent Anti-Stokes Raman Scattering
DLR	Deutsches Zentrum für Luft- und Raumfahrt
DNS	Direct Numerical Simulation
DNW	Deutsch Niederländische Windkanäle
DSMC	Direct Simulation Monte Carlo
EBF	Electron Beam Fluorescence

EFD	Experimental Fluid Dynamics
ENSI	École Nationale Supérieure d'Ingénieurs
ENSMA	École National Supérieure de Mécanique et d'Aérotechnique
ESA	European Space Agency
ETW	European Transonic Wind tunnel
GIE	Groupement d'Intérêt Économique (Economic Interest Group)
GIE S2A	GIE Souffleries Aérodynamiques et Aéroacoustiques
HEG	High Enthalpy Göttingen
HSM	High Speed Machining
IAT	Institut Aérotechnique (de Saint Cyr)
ICARE	Institut de Combustion Aérothermique Réactivité et Environnement (Institute of Combustion Aerothermal Reactivity and Environment)
IMFT	Institut de Mécanique des Fluides de Toulouse (Fluid Mechanics Institute of Toulouse)
ISAE	Institut Supérieur de l'Aéronautique et de l'Espace (National Higher French Institute of Aeronautics and Space)
ISL	Institut franco-allemand de recherches de Saint-Louis (French-German Research Institute of Saint-Louis)
IUSTI	Institut Universitaire des Systèmes Thermiques Industriels (Institute of Industrial Thermal Systems at the University of Marseille)
IWT	Icing Wind Tunnel
JAXA	Japan Aerospace Exploration Agency
LDV	Laser Doppler Velocimetry
LES	Large Eddy Simulation
LIF	Laser-Induced Fluorescence
LMFA	Laboratoire de Mécanique des Fluides et d'Acoustique (Laboratory of Fluid Transfer and Acoustics)
MDM	Model Deformation Measurement
MHD	Magneto-Hydro-Dynamique
NASA	National Aeronautics and Space Administration
ONERA	Office National d'Etudes et de Recherches Aérospatiales (French national aerospace research centre)
PGV	Planar Global Velocimetry
PIV	Particle Image Velocimetry
PLIF	Planar Laser-Induced Fluorescence
POD	Proper Orthogonal Decomposition
PPRIME	Institut PPRIME: Pôle Poitevin de Recherche pour l'Ingénieur en Mécanique, Matériaux et Énergétique (PPRIME Institute: Research and Engineering in Materials, Mechanics and Energetics)
PRISME	Pluridisciplinaire de Recherche en Ingénierie des Systèmes, Mécanique et Énergétique (Multi-field Research Laboratory in System Engineering, Mechanics and Energetics)

PROMETEE	PROgramme et Moyens d'Essais pour les Transports, l'Énergie et l'Environnement
PSP	Pressure-Sensitive Paint
PTV	Particle Tracking Velocimetry
RANS	Reynolds-Averaged Navier–Stokes
RMS	Root Mean Square
S2A	Souffleries Aéroacoustiques Automobiles (Full-Scale Aero-acoustic Wind Tunnels)
SBS	Spray Bar System
TPS	Thermal Protection System
TPS	Turbine Power Simulator
Tr-PIV	Time-resolved Particle Image Velocimetry
TSP	Temperature-Sensitive Paint
UAV	Unmanned Aerial Vehicle
UCAV	Unmanned Combat Air Vehicle
UPSP	Unsteady Pressure-Sensitive Paint
VG	Vortex Generator

Chapter 1
The Experimental Approach in Aerodynamic Design

1.1 Aerodynamics, What for?

Aerodynamics is the study of the flow of air around bodies, generally an aerial or ground vehicle, but also structures, turbomachine blades and rotors, wind turbines, or in more diverse cases such as bikes and other sports equipments. Indeed, any fluid in contact with a solid surface exerts a normal and tangential pressures, the integral effect of this pressure over the whole body generates a resultant force referred as the aerodynamic force. This resultant force is applied at a point known as the centre of pressure, which as a rule of thumb does not coincide with the centre of gravity of the body where the weight is exerted. The presence of these two points of action of the force induces an aerodynamic moment tending to rotate the body around its centre of gravity. A main goal in aerodynamics is to determine the components of this resultant force and their moments, the whole constituting the aerodynamic forces.

The aerodynamic resultant force is decomposed according to a system of axes linked to the relative direction of flight of the vehicle: the force can be decomposed into other components, the drag force which acts in the direction opposing the motion of the vehicle which must be compensated either by a propulsion unit (see Fig. 1.1), or a component of the weight in the opposite direction for example in the case of gliders.

Drag is the component that contributes to most of the energy consumption for propulsion. The component almost normal to the direction of flight or the oncoming wind is the lift. It requires very little or no energy at all and is used to compensate for weight and manoeuvres. A large majority of aerial and ground vehicles have a plane of symmetry, usually vertical in nominal operation. We can then break down the aerodynamic forces in this plane of symmetry, with the lift and drag lying on this plane and, the third component, perpendicular to the plane of symmetry of the vehicle, called the side force. It therefore defines the load carrying capacity of the aircraft referred as the payload (number of passengers, cargo mass). For automotive applications, the bump effect of the vehicle on the road could create lift, but this

© Springer Nature Switzerland AG 2020
B. Chanetz et al., *Experimental Aerodynamics*,
Springer Tracts in Mechanical Engineering,
https://doi.org/10.1007/978-3-030-35562-3_1

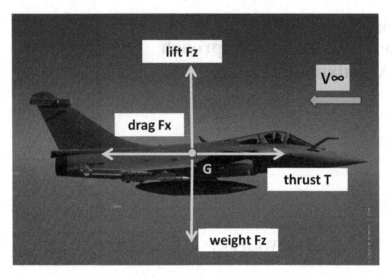

Fig. 1.1 Forces acting on a flying vehicle (© Dassault Aviation)

could affect grounding of the vehicle at high-speed and creates unnecessary drag. Here, we look for a zero or even negative lift when the goal is to turn at high speed, mainly in Formula 1 race.

The drag determines the propulsive force to exercise to keep the vehicle moving. This is an essential piece of information that determines fuel consumption. The side force is one of the components playing an important role in the stability of the vehicle. For example this could be the effect of a gust or a side wind on the handling of a car. Aerodynamic moments are also critical for the stability of the vehicle or structures. Earlier we introduced the notion of the centre of pressure variations which is the point of application of the variations of aerodynamic forces. The relative positions of the centre of pressure and the centre of gravity define the stability conditions.

In addition to controlling drag and lift, aerodynamic studies must also predict the local forces exerted on the body, especially the pressure, in order to calculate the deformations of the structure. The reciprocal influence between the structure, which by deforming modifies the aerodynamic field, and the flow which therefore imposes a varying load on the structure, is the origin of a fluid-structure interaction which can lead to severe conditions like flutter. A famous example is the failure of the suspension bridge of Tacoma which was triggered by a storm. This phenomenon, which is a concern for most aerodynamic applications including compressor or turbine blades, helicopter and wind turbine blades, brings us to the field of aeroelasticity.

Aeroacoustic deals with the interaction between aerodynamics and acoustics. The main focus in this field is on reducing the noise emitted by the turbulent boundary layer, the flow separations occurring on certain parts of the vehicle, the jets of the reactors or even those generated by propellers, helicopter rotors and wind turbines. Aerodynamic noise is also a concern for land vehicles, its level exceeding the rolling

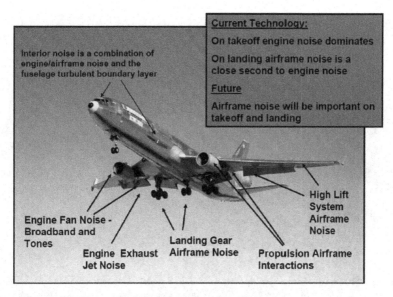

Fig. 1.2 Sources of noise on a transport aircraft (© LMFA)

noise beyond speeds of 300 km/h for trains and 120 km/h for cars. For aircraft, during the approach phase, whilst the engines are at low speed, the aerodynamic noise tends to become predominant, in particular that generated by landing gears. Figure 1.2 shows the main sources of noise on a transport aircraft. Convective and/or radiative heat transfer between the flow and the wall falls in the aerothermal or aerothermodynamic field. These aspects play an essential role in hypersonic flight where the vehicle is subjected to intense heating during a re-entry into the atmosphere. Conversely, a flow can be used to cool hot parts such as in a radiator or heat exchanger.

In military applications, the reflection of electromagnetic waves and the infrared radiation of fuselages and jet engines of combat aircraft and helicopter are used to detect their presence and lock them as target for or by guided missiles. Stealth, that is to say the ability of a combat vehicle to escape these means of detection, imposes very specific shapes, which forces us to reconsider their aerodynamics (Fig. 1.3 shows an image of a combat drone or UCAV).

The impact of aerodynamics is also considerable in the quest for energy savings, which would help the reduction of noxious gas emission otherwise affecting our environment, or simply the noise when the vehicle passes by. This encourages the design of optimised vehicles while focusing on reducing the drag in particular.

It is very useful to define non-dimensional aerodynamic coefficients which are the aerodynamic forces normalised by the product of the dynamic pressure of the freestream flow ($1/2\ \rho V^2$) and the reference area of the vehicle. For an aircraft this reference surface is generally the planform area of the wing. For a land vehicle or a bluff body, the frontal area is used instead. By way of illustration, Table 1.1 gives the coefficients of drag (C_D), lift (C_L) and the lift-to-drag ratio (L/D) which

Fig. 1.3 Unmanned combat aircraft vehicle, nEUROn (© Dassault Aviation)

Table 1.1 Aerodynamic coefficients of civil transport aircraft

	Cruise			Take off		
	C_L	C_D	L/D	C_L	C_D	L/D
Subsonic transport	0.50	0.027	18.5	1.50	0.130	11.5
Supersonic transport	0.12	0.012	10.0	0.40	0.045	8.9

characterises the flying quality of civil transport aircraft. At takeoff, high lift is sought at the expense of drag, which is greatly increased; this reduces the lift-to-drag ratio. In cruise flight, the drag is decreased without sacrificing the lift while operating near the point of maximum lift-to-drag ratio.

Table 1.2 shows the drag coefficient, C_D, of ground vehicles where the drag, D, is normalised by the frontal area of the vehicle.

Figure 1.4 shows the evolution of the power dissipated by a road vehicle as a function of speed on a horizontal road. The aerodynamic contribution appears to be significantly larger than the contribution due to friction between the tyre and the ground. From the figure on the right, the contribution from aerodynamics quickly becomes paramount: at 50 km/h, 50% of losses are due to aerodynamic drag already, and 85% at 130 km/h.

In rail transport, aerodynamics plays a key role in bringing high-speed trains into service. Thus for these trains running at 300 km/h, the aerodynamic drag represents 80% of the resistance and the progression to the record speed of 575 km/h reached in April 2, 2007 saw this drag increasing towards 90%. The significance of this

Table 1.2 Drag coefficient of land vehicles

	Frontal area S (m^2)	C_D	$S \times C_D$ (m^2)
Sedan car	1.8	0.30	0.54
Minivan	2.6	0.50	1.30
Trailer	9.0	0.90	8.10
Formula 1	1.6	0.90	1.44
Moto–bike	0.7	0.90	0.63
Cyclist	0.5	1.00	0.50

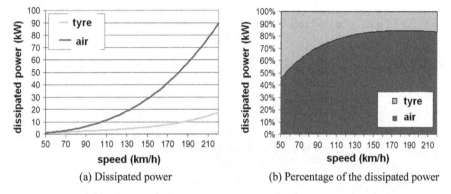

(a) Dissipated power   (b) Percentage of the dissipated power

Fig. 1.4 Power dissipated by an automobile rolling at constant velocity on an horizontal road

resistance force leads to careful consideration of the aerodynamics of trains and identification of the main sources of drag, for example from the front carriage, bogies, spaces between cars, pantograph, etc. Figure 1.5 shows a comparison of the running resistance of high-speed trains with that of a conventional Corail train (which of course does not travel at 300 km/h); the lower resistance is achieved by a more careful aerodynamic design.

In general, aerodynamics is perceived as a science where the movement of gaseous fluids is studied. It is of interest to a diverse range of other fields such as ventilation and air conditioning (we speak of aeraulics), weather forecasting, buildings and structures in civil engineering applications (see Fig. 1.6), metal casting and various industrial processes in manufacturing; but not limited to biomedical applications, cardiac-vascular, and resuscitation.

In the field of propulsion and energy production (combustion engines for cars, jet engines for aircraft, turbogenerators for thermal power plants, wind turbines), aerodynamics also plays a key role and in some machines implementing fluids that can reach supersonic speeds.

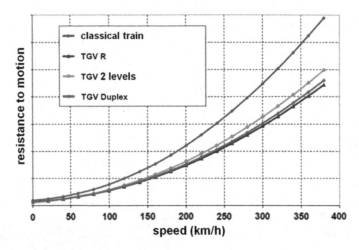

Fig. 1.5 Comparison of the running resistance for different types of trains (© SNCF)

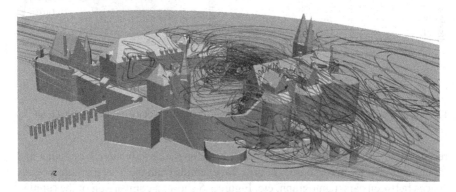

Fig. 1.6 Study of wind effect on the Château des Ducs de Bretagne (© CSTB)

1.2   Review of the Theory

The Navier-Stokes equations, which formalise the mathematical representation of
the aerodynamic behaviour of moving bodies in a fluid, will not be developed here.
Resulting from the application of the principles of mechanics and thermodynamics,
these equations include evolutionary terms for mass, motion (application of Newton's
law) and total energy (sum of internal energy and kinetic energy) supplemented by
laws reflecting the effects of viscosity and thermal conductivity.

These equations do not have a simple analytical solution in general, but it is
possible to approach it by a combination of locally simpler functions mainly based
on some assumptions valid for engineering applications. This is the goal of computer
fluid modelling or Computational Fluid Dynamics (CFD). These functions must
respect the Navier-Stokes equations in a number of control points in the fluid, so

Fig. 1.7 Mesh for the calculation of the flow past a business jet by a finite element method (© Dassault Aviation)

since the flow is more complex due to the presence of both small and large length scales, the number of points usually reaches several millions and occasionally several billions. The functions must also respect boundary conditions which represent the solid wall of the object and the state of the fluid very far from the object usually referred as the far field. It is thus necessary to mesh the space around the vehicle, which is a complex and a potentially time consuming operation. An example is given in Fig. 1.7 which represents the mesh on the body and a section in the plane of symmetry of a business jet.

Initial and boundary conditions are fundamental for the appropriate solution of the Navier-Stokes equations. Thus, the computational domain has physical boundaries, sometimes well defined as the wall of an aircraft in contact with the atmosphere, but also permeable surfaces could be mathematically modelled within a defined computational domain. On the solid walls, the aerodynamic field satisfies a dynamic condition and a thermal condition: first, the no-slip condition, which models the adhesion of air to the wall and the second relates to the prescribed temperature and/or heat flow. Boundary conditions on permeable boundaries must be carefully formulated to avoid introducing spurious nonphysical effects.

In the very low-pressure flows experienced by high-altitude vehicles during atmospheric re-entry, air can no longer be considered as a continuous medium, the inter-molecule distances or free mean molecular path starts becoming comparable to the length scales of the vehicle. It is then necessary to abandon a continuum modelling, formalised by the Navier-Stokes equations, to a discrete approach like Direct Simulation Monte-Carlo (DSMC) method. The so-called rarefaction effects are quantified by the Knudsen number (see Sect. 2.4). To describe an intermediate situation, between continuous and rarefied regimes, one can use the Navier-Stokes equations by admitting a certain slip of the fluid on the walls. This point will not be discussed here.

Table 1.3 Mean molecular free path in the atmosphere

Altitude (km)	λ (m)
20	10^{-6}
70	10^{-3}
110	1
150	10

At "aeronautical" altitudes, that is to say, less than about 20 km, the mean molecular free path is less than one micron, so much smaller than the size of the vehicle, which is of the order of several meters (see Table 1.3), this justifies the assumption of a fluid as a continuum and the validity of the Navier-Stokes equations. At 70 km, the mean free path becomes of the order of a millimetre, which is not very small vis-à-vis the re-entry body, but still small enough to consider the air as a continuous medium, except perhaps in some areas where the flow expands strongly. Beyond 100 km, the average mean free path becomes comparable to the length of the vehicle and the assumption of continuous medium for the air is again invalid.

From dimensional analysis, the physical quantities of the continuous model are defined by four quantities, being a unit of length adapted, the density, ρ, temperature, T, and the velocity, V, for a given reference state. To measure the relative weights of the different terms of the equations, it is usual to introduce additional reference quantities. For example, if we are interested in the equation of motion that reflects a balance of inertial forces, pressure and viscosity, it is convenient to introduce two new reference quantities: one for the pressure ρV^2 or density, another for the viscosity term ρVL. By introducing these two additional terms into the momentum equation two dimensionless numbers are obtained and are very important in aerodynamics. These are:

- the Mach number, $M = V/a$, ratio of the velocity V to the local speed of sound, a and,
- the Reynolds number, $Re = \rho VL/\mu$, ratio of the inertial forces to the viscosity forces where μ is the molecular or dynamic viscosity of the fluid.

These two numbers are critical dynamic similarity parameters for the extrapolation of the actual wind tunnel conditions to flight conditions or vice versa (see Sect. 2.4).

As shown in Fig. 1.8, the aerodynamics of aircraft and land vehicles is characterised by very high values of Reynolds number (several million) reflecting the low viscosity of the air.

Thus, for an Airbus A380, the Reynolds number at cruise, based on the average chord of the wing, is approximately 70 million. Since this is a large value, some terms of the Navier-Stokes equations become negligible and we can consider that the flow in the entirety is almost a "perfect" fluid. By neglecting viscosity and thermal conductivity; the flow can be modelled by a simplified set of equations called the Euler equations. The viscous effects (friction, thermal conduction) are then modelled in confined regions (boundary layers and/or mixing layers, wakes) which have very large variations in velocity and/or temperature, compensating for the very low values of

Fig. 1.8 Reynolds numbers based on the length L of the vehicle for a kinematic viscosity equal to 15×10^{-6} m^2/s

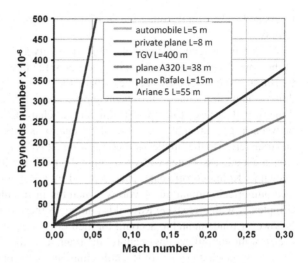

viscosity and/or thermal conductivity. There are two main types of confined regions depending on whether the velocity varies along the main direction of the flow or in a direction normal to the wall. The first case corresponds to the shock region or layer whose thickness varies inversely with Reynolds number, the second, the boundary layers at the walls and wakes whose thickness varies as the inverse of the square roots (laminar case) or to a power (turbulent case) of the Reynolds number. At an altitude of 20 km if the Mach number of the vehicle is 10, the thickness of the shock will be of the order of mean molecular free path, therefore very small for a body whose length is several meters. However, at 150 km the mean free path is approximately 10 m, for the same Mach number: thence the thickness of the shock is of the order of the size of the vehicle.

Euler's equations are first-order equations that model a physics corresponding to non-linear propagation phenomena without diffusion (transport by velocity and by acoustic waves). When the flow is subsonic, the perturbations by the flow propagate both downstream and upstream of the source of perturbation in what is usually referred as the elliptical propagation. For instance an observer on the ground perceives the noise of a plane in subsonic flight when it approaches or when it moves away. When the flow is supersonic, the perturbations propagate only downstream in the hyperbolic condition; here the same observer does not perceive the sound of an airplane coming towards him in supersonic flight. The Euler equations have the particularity of allowing discontinuous solutions: shock waves and contact discontinuities (slip lines). A fluid particle passing through a shock sees its entropy and pressure increase. For a stationary shock normal to the direction of flow, the velocity is supersonic upstream of the shock and subsonic downstream. Over the slip line, the fluid particles slide on both sides of the discontinuity, the entropy being able to undergo an arbitrary jump while the pressure remains continuous. Contact discontinuities and shocks are the limiting forms; on one hand we have the wall boundary layers and wakes, on the other the shock layers when the Reynolds number tends to infinity.

At very high Reynolds number further difficulty arises due to the unstable nature of the flow. As the Reynolds number increases, the flow over a surface can undergo transition from a laminar regime, where the flow is stationary and almost two-dimensional, to a chaotic turbulent regime characterised by unsteady and three-dimensional vortex structures. Turbulence is characterised as an energy cascade between large and smaller eddies, where energy is extracted from large structures and transferred to the smallest structures where it is dissipated into heat. We thus pass from a scale L, associated with a very high Reynolds number Re, to the smallest scale known as Kolmogoroff scale, characterised by a Reynolds number of about unity. The numerical simulation of vortex structures in turbulent flows is of increasing complexity with increasing Reynolds number as these structures are even smaller. It is however fundamental to account for their effect on the characteristics of the flow, even if some simplifications or approximations are made while modelling turbulence.

For the very high values of Reynolds number, of the order of one million or more, the Reynolds Averaged Navier-Stokes (RANS) statistical approach is by far the most used method for most practical aerodynamic studies. When the Reynolds number is smaller, of the order of one hundred thousand, the scale of the turbulent structures increases, we can describe quite precisely the largest of them thanks to the methods called LES (Large Eddy Simulation). This approach is based on a separation between large scales, described by numerical simulation, and small unresolved scales to be modelled. The LES methods require a great refinement of the mesh to describe the attached boundary layers. An intermediate approach called Detached Eddy Simulation (DES) has been developed in the LES method which is activated only when the flow separates. The very rapid progress of the computing resources (Moore's law!) has led to these tools to be used on full aircraft configuration in the aerodynamic design offices and will probably be in common use in less than a decade.

Direct Numerical Simulation (DNS) does not pose a problem of modelling since it solves the full instantaneous Navier-Stokes equations, assuming that they represent the turbulent motion down to the smallest scales. In practice, this constraint leads to meshes of very high density and consequently to very long computation times which increases rapidly with the Reynolds number. Their applications are currently limited to the simulation of fundamental aerodynamic problems. Despite the advances in computational power it is still not a viable tool for applied industrial problems.

Another approach, currently limited to incompressible flows, is the Lattice Boltzmann Method (LBM), introduced in the mid-1980s. This technique involves modelling the fluid as a set of particles while expressing all the physical quantities characterising their trajectories, these quantities are length, speed and time. The particles are free to move on a lattice or regular network of points called nodes, displacements being inferred from a two-step protocol. The first associates with each node a distribution function for the velocities; the second step is to model the collisions between particles. Currently, this approach is mostly used in the automotive industry.

1.3   Limitations and Constraints of Numerical Methods

Until recently, the validation of numerical methods was mainly done by comparing calculated results with overall aerodynamic forces measurements and properties measured at the wall, essentially pressure. For many studies this type of comparison was sufficient. Such are the "traditional" methods, empirical or based on a simplified modelling approach, predicting only wall properties such as pressure, skin friction, heat flux and the overall performance of the vehicle. These methods could also give an idea of the overall flow-field, for example, predicting the size of a separated region and the location of a point of separation, but this information was considered more or less qualitative. The scope of flow prediction has changed with the advent of theoretical models based on solving the Navier-Stokes equations (see above) or lower-order models formulated from simplification of these equations. It is clear that this approach is the only one capable of calculating complex flows containing shock waves and expansion fans, detached regions and vortex structures, where the dissipative regions being turbulent in almost all practical situations. Not only the wall properties are calculated, but other quantities in the flow-field including the average velocity and the complete turbulent field. However, in its current state the Navier-Stokes approach is still far from being free from criticism, many difficulties remains in the numerical methods and in the physical modelling, especially for the prediction of the full turbulence field. There is therefore a strong need to validate codes prior to their routine use for design purposes.

Although prediction of wall properties remains an essential goal for most computational methods, since lift, drag and in some cases wall temperature are the quantities of greatest practical interest, it soon became clear that the comparison with wall properties was insufficient to validate most of the advanced predictive methods. In general, the Navier-Stokes codes give a faithful and impressive image of the aerodynamic field structure. However, a closer look at the results shows that the situation is far from entirely satisfactory. Thus, it can be seen that a fairly good prediction of the wall pressure can coexist with a poor quantitative description of the velocity field. Frequently, the extent of a separated region is underestimated, sometimes considerably. In addition, the turbulent quantities are poorly predicted, especially if the flow is in great part separated. These discrepancies may make the validity of the method suspect, since they reveal a certain deficiency either in their numerical scheme, or in the turbulence model, or both. Conversely, a satisfactory prediction of the field can be accompanied by significant disagreements in the calculation of the surface properties affecting the transfer quantities, such as skin friction and heat transfer.

In some applications, knowledge of the external far field itself is of primary interest, as in the case of the infrared signature where the detailed description of the hot propellant jet, with a precise location of the Mach disks, is essential. Air pollution studies require good prediction of the flow field to allow for a proper assessment of chemical reactions and species concentration as well as pollutant dispersion. The same is true for the prediction of vehicle noise, where the prediction of the aerodynamic noise level and the analysis of its origin being an important issue for aerodynamic installations (see Sect. 3.2). In a lesson learnt from the sonic boom of the Concorde, for a supersonic aircraft, the origin and control of shock waves that propagate to the ground is a critical issue.

1.4 Some Constraints for Wind Tunnel Test

There is a wide range of experimental means to carry out measurements required to validate the numerical techniques mentioned above, but their degree of reliability, ease of use or accuracy are extremely variable. In practice, it is often very challenging to perform several types of measurements on the same model or in the same facility or laboratory. Indeed, the use of sophisticated techniques required to perform these measurements is so complex that it may necessitate highly specialised teams of experimentalists working on specific installations. Thus, for the purpose of validating the numerical models, it is often necessary to rely on several experimental campaigns, where first the aerodynamic forces, pressure and temperature distributions are measured, followed by more complicated measurement of the average and turbulent flow-fields from another technique and possibly the density from a third source in the case of compressible flows.

Nevertheless, advances in measurement techniques over the last 40 years, including the advent of laser-based optical methods, have made a real breakthrough in our ability to analyse complex flows containing shock waves, strong expansions, shear layers, vortex organisations and recirculation regions.

Running experiments in a wind tunnel (or any similar installation) is also a simulation, here similar environment in which the vehicle will operate is recreated, the model being tested being often much smaller than the full scale vehicle. As mentioned above it is of great importance to reproduce the viscous effects quantified by the Reynolds number, which should be identical for the small scale model and the real vehicle. Other similarity parameters such as the Mach number for compressibility effects, the Prandtl number for heat conduction, the Lewis number for mixing species, the Knudsen number for rarefaction effects, must be conserved as well (see Sect. 2.4). In hypersonic studies, it is recommended to use a gas with the same composition and thermodynamic properties as the actual gas. Not all characteristic numbers can be reproduced simultaneously and various facilities have to be used to simulate the different flow conditions encountered by the vehicle, some conditions being impossible to reproduce in ground facilities.

During most wind tunnel tests particular attention is paid to the Reynolds and Mach numbers, the gas (in this case air) being the same in the practical application and in the wind tunnel. However, during validation studies focusing on the physics of complex aerodynamic phenomena, the Reynolds number is not as critical as it is proclaimed. Indeed, for a well-established turbulent regime, the characteristics of a flow are almost independent of the Reynolds number. To a large extent, the main influencing parameters are boundary-layer properties just prior to separation (velocity

and turbulence distributions). The Reynolds number influences the upstream development of the boundary layer according to a history effect, not by a local effect.
There are many observations in favour of this quasi-independence with regard to the
Reynolds number: formation of vortices on a delta wing, vortex breakdown, base
flows, shock-induced separation, cavity flow to list a few. The important thing is to
achieve an established turbulent regime for the upstream flow and to provide a precise definition of the boundary layer. Further examination of the physics shows that
the thickness of the boundary layer is the appropriate length scale if the extension of
the separated domain is small. On the other hand, if the separation is extended, the
characteristic scale becomes the distance from the point of separation, the influence
of the initial boundary layer being quickly forgotten.

1.5 Deformation of Models

Another important problem for representativeness of tests or calculations is geometric
conformity. Small scale models, most often representative of the overall shape, cannot
reproduce the levels of detail such as surface defects (screws, rivets, structural joints,
etc.) which can, themselves, vary between two versions of the same production
vehicle. The general forms are themselves difficult to define, for instance the wing
tips of an Airbus A380 deflect by several metres in flight, significantly altering the
flow around the wings and this alters trimming characteristics of the aircraft. The
wind tunnel models are also not infinitely rigid and are normally subjected to high
aerodynamic loading due to the large dynamic pressure required to get closer to
the flight Reynolds numbers. Therefore they are bound to deflect and deform. Thus
the extrapolation of wind tunnel measurements to the real operational condition
is a complex problem even beyond the complexity related in achieving dynamic
similarity. During wind tunnel testing, model deformation measurement techniques
have been developed and are used to transpose the results to reference geometry
(see Sect. 9.6). In flight, the precise identification of the shape of the aerodynamic
surface in different flight conditions remains a difficult problem which is generally
only approached by aeroelastic modelling. For CFD calculations, the geometry is
known and controlled but rarely corresponds to the actual geometry in flight, which
itself varies according to the flight conditions.

It is theoretically possible to design flexible models that deform in the same
way as the aerodynamic surface of the vehicle in operation. In practice this is not
feasible on complex models, but can be done on simplified models to study critical
phenomena of aeroelastic coupling such as flutter and force response. These models
are expensive and require very specific testing procedures. These difficulties are
more rarely encountered in the testing of ground vehicles where most often full scale
prototypes are tested, except while testing very large vehicles or structures where
once again model scaling is required, for instance trains and other very large size
vehicles, including ships or buildings.

1.6   Industrial Aerodynamics Testing: Combining Tests and Numerical Simulation

The turn of the century has seen tremendous progress in numerical simulation methods as a result of the even more staggering improvement in computing resources. RANS methods are used extensively in industry regardless of the complexity of the geometries. The scope of these methods is well defined and gives the designer the opportunity to carry out a considerable amount of the design work by numerical simulation mainly in the automobile, aircraft, propulsion and turbomachinery industry. However, this approach has its limits: the zones of strong interaction, of separation on a curved surface or of massively detached flow are notoriously difficult to predict with precision. In addition, since the automotive and aerospace industries are highly competitive and will innovate through concepts for hypersonic flight, active flow control or radically new geometrical configurations, this may involve physical phenomena for which the tools have been inadequately or not validated at all. It will then be necessary to revert to wind tunnel tests at the design stage in order to validate the numerical tools to handle the new phenomena or to combine numerical and experimental approaches (see Chap. 13).

If we take the example of the flight of an air-breathing hypersonic vehicle, the propulsion system is so integrated that it is difficult to determine its air-propulsive balance by separating the propulsion on one side and the drag on the other. Here, the designer is faced with a situation where not only it is difficult to carry out ground tests, that are fully representative of the real flight, but also the validation of the calculation means is flawed as the airframe has to be decoupled with the propulsion system. It is therefore essential to develop an approach that closely links numerical simulation and testing at all stages of design. The engineer will have to implement this approach on a more fundamental level as well to account for issues related to boundary layer transition at very high altitude flight and high Mach numbers (quiet wind tunnel tests to validate the transition mechanism are presented in Sect. 5.4), rather than just the global aspects such as aero-propulsive assessment. This can be evaluated either by means of calculations coupling the external aerodynamics, the internal aerodynamics, the combustion chamber and the nozzle, or by tests in free jet installations, often partially representative of the flight conditions.

For a more conventional vehicle, the wind tunnel remains a very effective and reliable means of characterisation. It is indeed possible to acquire within a reasonable time a very large amount of data that would be difficult to obtain by numerical calculation. However, once again, the experimental approach suffers from approximations due to support effects, wall effects, Reynolds number corrections and model deformations that are not representative of the flight, to mention a few. We would like to emphasise on the fact that neither experimental approach nor numerical simulation are means to an end, they are in fact complimentary to each other and if employed rationally it will get us as close as possible to reality (see Chap. 13).

1.7   Flight Tests

1.7.1   Flight Test Beds

Here, we do not consider flight tests intended to test the performance of a new aircraft and to define its flight range prior to delivery to the airliner, but experiments where the aircraft is used as a flying test bed to carry out experiments in conditions closer to reality rather than those achievable in a constrained wind tunnel. This has the benefit of reproducing the Reynolds number and the very low freestream turbulence which is primordial for testing the techniques for drag reduction, such as natural or hybrid laminar flow control. We will see in Chap. 2 and Sect. 5.4 that the flows produced in the wind tunnels are affected by environmental or freestream disturbances which are very difficult to eliminate. This can lead to unreliable experimental results due to premature laminar to turbulent transition and also potential misinterpretation of the result during the design phase culminating in more severe consequences.

Below are examples of two studies employing an aircraft as a flying test bed. They are both for the study of transition to turbulence on wings a key motivation for this study being skin friction drag reduction in the search for energy savings. The first, shown in Fig. 1.9, was a test carried out on a Falcon 7X business jet where the laminar to turbulent transition was visualised on the horizontal stabiliser by an infrared camera mounted on the fin (see Fig. 1.9a). The resulting infrared image shows the laminar regions that appear in green in Fig. 1.9b.

Maintaining a laminar state on an actual aircraft wing can be compromised by the existence of surface excrescences and defects, particularly at the junction of leading edge slats or anti-icing devices. The second example here is a study of the influence of these defects on transition, performed by William Saric and co-workers at Texas A&M University on a swept wing section mounted vertically under the wing of a Twin Cessna O-2A Skymaster as shown in Figs. 1.10 and 1.11.

(a) IR camera installation   (b) Infrared image of the horizontal stabiliser

Fig. 1.9 Detection of laminar to turbulent transition on the rear horizontal stabiliser of the Falcon 7X (© Dassault Aviation)

Fig. 1.10   The twin-engine Cessna O-2A Skymaster (© A&M University)

(a) Model attached to the wing lower surface   (b) Infrared image showing the transition

Fig. 1.11   Model installation for the study of transition on a wing portion (© A&M University)

The wing model is equipped with an actuator that can displace the leading section thus creating a forward or backward facing step in the upstream region of the wing. The effect of these steps on transition was detected by infrared thermography (see Fig. 1.11b), the laminar regions appearing in yellow. Many similar flight experiments have been performed on various devices at Mach numbers ranging from subsonic to supersonic speeds, but these experiments are very expensive; they are mainly conducted to validate a concept that has showed potential benefits during wind tunnel test campaigns.

1.7.2   Catapulted Flight Test

Another type of flight test is by catapulting a model and studying its behaviour in free flight until impact with the ground. One of these very old techniques, traditionally used to study the stability of aircraft, is implemented at the ONERA research centre in Lille. This research centre has a laboratory where these kinds of studies

(a) Aerial view of the building: length 90 m,   (b) Flight of a catapulted model

width and height 20 m

Fig. 1.12   The free flight laboratory of the ONERA Lille centre (© ONERA)

can be conducted at relatively low costs and in a controlled environment, whereby research in flight modelling and aircraft flight mechanics as well as more fundamental aerodynamic tests can be carried out (see Fig. 1.12a).

The laboratory also offers the possibility of conducting in-flight demonstrations to validate a concept and evaluate its performance. The test method consists in catapulting the model of the aircraft to be studied in free flight, at a desired speed and attitude. The tests are carried out by catapulting under the conditions of a trimmed flight. The model is launched by a carriage that slides on the rail of the catapult (see Fig. 1.12b). The model sits on metal tips with minimal area of contact on the carriage which is driven at controlled speed. The braking of the carriage at the end of the ramp releases the model due to the inertial effect and the model is launched in free flight under the equilibrium conditions previously defined. During the flight, various types of conditions can be imposed on the model, some of them are: steering, vertical or lateral wind gust, ground effects, jet impingement, etc., and the behaviour under these flight conditions can be studied. The model is recovered at the end of flight in a collector designed to preserve its integrity, including the empennages and nacelles or other additional parts.

This technique was used to characterise the wake vortices of a large commercial aircraft. Indeed, such vortices, which remain active over a considerable distance downstream of the leading aircraft (several kilometres), are hazardous for another aircraft operating within the vicinity of these large vortical perturbations. Hence the imposition of minimum safety distances between two successive take-offs which tends to slow down airport traffic. Because of their long persistence, the study of these vortices in the wind tunnel would require unpractical length of test section or the use of small models significantly compromising the Reynolds number. In the experiments performed at ONERA in Lille, the model launched by a catapult passes through a cloud of smoke illuminated by a high intensity light sheet. As shown in Fig. 1.13, by taking snap shots at different time intervals after the passage of the

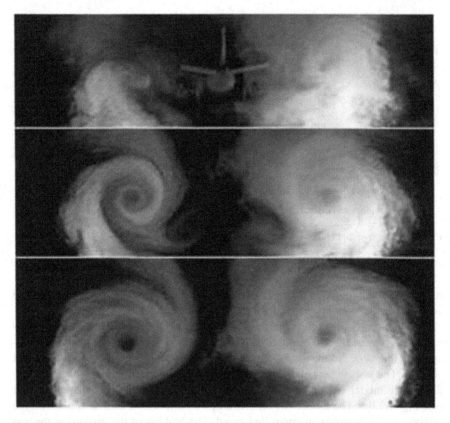

Fig. 1.13 Wake vortices of a transport aircraft model in the free flight (© ONERA)

model the evolution of the wake vortex in time can be studied as well as the growth
and decay distance downstream of the model. Today, this installation is mainly used
for the development of stabilisation technologies for small UAVs (Unmanned Aerial
Vehicles) subject to freestream disturbances.

1.7.3 Aeroballistics Flight Test

To avoid interference related to the support of the model, free flight experiments can
be conducted using aeroballistics launches, allowing for improved dynamic stability
of the body. Less than twenty years ago, this technique was performed in an aeroballistics range or firing corridor. It consisted of firing the model with a cannon
and recording the image of its passage on a photographic plate using shadowgraph
technique with light source consisting of a spark generator emitting a very short flash
of light (see Fig. 1.14). This technique has evolved towards the use of a very high

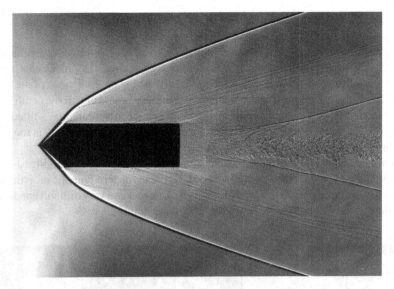

Fig. 1.14 Projectile in free flight in the ISL aeroballistics range (© ISL)

resolution digital camera. However, the small number of stations along the firing axis is a limiting factor for obtaining the aerodynamic force coefficients.

At present, more modern free flight techniques make it possible to obtain the evolution of the aerodynamic coefficients over the entire trajectory. The model is instrumented with magnetic sensors, accelerometers, gyrometres and other electronic sensor management and signal transmission units and is normally launched by a gun. The models being more and more complex, their launch requires that they have to be integrated with a sabot. Still, the on-board electronics must withstand the acceleration of the projectile that can reach 50,000 g (g: acceleration of gravity). The SIBREF facility (Soft In-Bore REcovery Facility) of the French-German Research Institute of Saint-Louis (ISL) allows all components embedded in a model to be tested at accelerations ranging from 5000 to 60,000 g. The signals measured by the sensors during flight tests are processed by an inverse ballistic code with 6 degrees of freedom to determine the aerodynamic coefficients.

An innovative technique developed at the ISL consists in determining the position and the attitude of a model in free flight from video images acquired at a high frame rate. For this, two optical tracking systems (trackers) are located on either side of the firing line, the difficulty residing in the adjustment of the speed of the motorised mirror of each tracker to the speed of the model. The position and attitude of the latter are determined from a mathematical modelling of each optical tracking system and an image processing algorithm to isolate the model in the image. The three-dimensional position and the attitude of the model along its trajectory can then be compared to those obtained from the onboard sensor technique. These tests are performed on the ISL proving ground located north of Mulhouse, the shots being carried out over a distance at most equal to 1000 m with powder cannons (with rifled or smooth bores) of calibre ranging from 20 to 105 mm. The launch Mach number, depending on the configurations studied is between 0.6 and 6.0.

Figure 1.15 shows an aerial view of one of the firing sites as well as the measuring devices. The speed of the model is measured by a continuous Doppler radar placed at the rear of the gun. Both trackers (A and B) are visible on both sides of the firing line. Due to the high cost of instrumentation of the sensor and electronics embedded in the model, it is recovered gently by means of one or more containers filled with rags or tires. If the model cannot be retrieved because its kinetic energy is too high, it is brought to rest upon impact with the sand bay.

Since this book is being dedicated to the means of ground tests, the flight tests were only briefly mentioned although they constitute a means of aerodynamic investigation still in use, offering possibilities of simulations of the real and very difficult conditions of a flow, if not impossible, to reproduce in a wind tunnel.

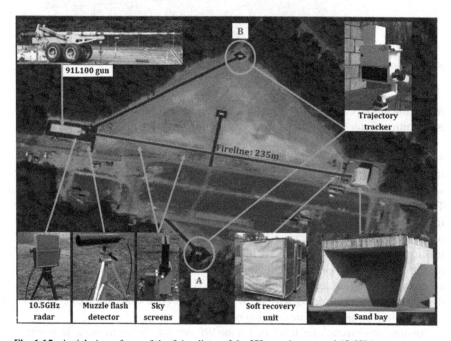

Fig. 1.15 Aerial view of one of the firing lines of the ISL proving ground (© ISL)

1.8 Simulated Altitude Test Cells

1.8.1 Impact of Altitude

Wind tunnels are useful to study aerodynamic phenomena under atmospheric pressure that is to say at ground level. However the difference in atmospheric conditions at varying altitude may have significant effects, especially on the operation of air-breathing engines and on the risk of ice accretion.

The category of air-breathing engines for both manned and Unmanned Aerial Vehicle (UAV) includes turbojets and turboprops, helicopter turboshafts and missile turbojets (see Fig. 1.16). They are subject to atmospheric conditions at varying altitude during flight: the higher the altitude, the lower the ambient static pressure and temperature. This has an impact on the working of gas turbine engines. For instance, if the mass flow rate of the engine is lower at altitude due to lower pressure, in-flight re-lighting is more difficult and so on.

Altitude has a significant influence on ice accretion as well. When an aircraft is flying through clouds ice may accrete on probes (Pitot probes for example), the leading edge of wings and/or at the engine intakes. Such ice accretion leads to an increased risk of system failures or severe consequences. Indeed ice accretion on a

Fig. 1.16 Icing test on a helicopter intake at simulated altitude conditions—Test cell R6 (© DGA Aero-engine Testing)

probe affects measurements, which may lead to pilot misinterpretation. Whilst, ice accretion on wings modifies the airflow around the wings: advancing stall angle, increasing drag and reducing manoeuvrability. The shape of ice formed on an engine intake modifies the airflow entering the engine and thus the engine performance. Furthermore, ice blocks formed on the engine intake may detach and impact the compressor blades, leading to damage and possible engine failure.

Altitude test facilities make it possible to carry out tests under the same conditions to that at a given altitude, including simulating icing conditions.

1.8.2   How Does an Altitude Test Cell Work?

The DGA (DGA is the French Defence Procurement Agency) Aero-engine Testing branch is specialised in testing under simulated altitude conditions. Since 1946, it has been designing and carrying out test campaigns on air-breathing aero-engines, their components, assemblies and sub-assemblies and associated equipment, under simulated flight and icing conditions. The operation of these facilities will be elaborated below.

The test article is normally mounted in a cylindrical test cell which is linked upstream and downstream to a network of pipes, valves, air supply and exhaust facilities, as shown on Fig. 1.17.

This network (2 km long, up to 3.2 m in diameter) enables accurate regulation of the conditions inside the test bench. The upstream located:

– compressors and turbines regulate the pressure,
– chillers and heaters regulate the temperature,

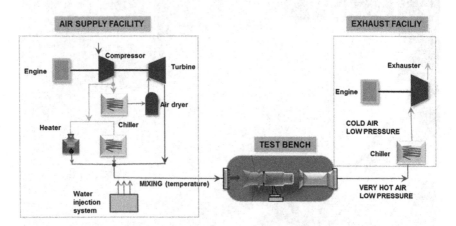

Fig. 1.17   Schematic diagram of a simulated altitude test bed (© DGA Aero-engine Testing)

Fig. 1.18 The air conditioning system for air supply (© DGA Aero-engine Testing)

– an air dryer and a water injection system regulate humidity, the main function of
 the air dryer being to remove the moisture of the air upstream of the turbine to
 avoid damaging its blades.

Downstream, cooling systems such as chillers, water injection systems, water
cooling jacketed pipes, channels where pipes are immersed are used to cool the working gases (see Fig. 1.18). Compressors (exhausters) enable regulation of pressure and
release of air back into the atmosphere.

The air supply and exhaust system enable the recreation of the same temperature,
pressure, humidity and mass flow rate as the corresponding altitude and aircraft
velocity required by the customer. They can provide:

– a mass flow rate up to 150 kg/s,
– a pressure from quasi vacuum to 22 bar,
– a temperature from −70 to 520 °C.

The above-mentioned range of values is given for all the test-beds of DGA Aero-
engine Testing facility, including combustion chamber test rigs. Conditions at the
outlet of the compressor are reproduced upstream the combustion chamber, where
the pressure can reach, 22 bar and temperature, 520 °C.

1.8.3   Benefits of Simulated Altitude Tests in Addition to Ground

There are several advantages to carrying out simulated altitude tests rather than in-flight tests:

– cost reduction,
– risk mitigation,
– control of the altitude and atmospheric conditions, especially for icing tests (in-flight tests are dependent of the weather),
– repeatability of test conditions,
– larger number of measurements (around 1000),
– ability to reproduce the conditions under which a failure took place.

In addition to simulated altitude tests, ground testing is also performed at DGA Aero-engine Testing.

Simulated altitude test cells can simulate altitudes of 0 m and even down to – 610 m, to represent some terrestrial areas which are situated below sea level. This allows the test cells to perform ground and altitude tests on the same test-bed, to optimise tests (one test bench instead of two) and avoid repeatability issues arising from the use of different setups.

The DGA Aero-engine Testing branch ground test-beds are listed below.

– The T_0 test bed which can perform engine ageing tests.
– The PAG is a wind tunnel for icing tests on probes and small equipment (see Fig. 1.19); it can perform tests more cost effectively than a simulated altitude test cell used at ground conditions.
– The GIV test bench can carry out icing tests on fuel systems.
– The K9 test bench is used to test combustion chambers.

Fig. 1.19 PAG test bed for
icing tests (© DGA
Aero-engine Testing)

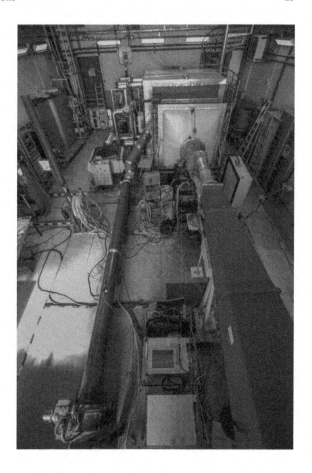

Chapter 2
Wind Tunnels and Other Aerodynamic Test Facilities

2.1 Background of Wind Tunnels

2.1.1 Wind Tunnel Principle

The wind tunnel is a means of studying and understanding the behaviour of an aerial or ground vehicle by performing an experiment, usually on a scaled model. The direct force measurements (mainly the lift, drag, side force and moments) can be extrapolated to the actual vehicle if dynamic similarities are satisfied. The wind tunnel, or other means of ground testing, also allows a detailed characterisation of the flow while measuring the wall pressure, skin friction, velocity and turbulence fields, etc. Through the implementation of appropriate measurement techniques, these quantities in particular allow very detailed validation of the numerical and analytical methods. The wind tunnel experiment allows the analysis of certain critical phenomena occurring at extreme conditions, such as massive separation, unsteadiness, buffeting, flutter and many more. These experimental facilities also allow detailed study of local phenomena that are detrimental for the proper operation or performance of the overall system. Some of these phenomena are shock-wave/boundary-layer interactions, the development of mixing zones, vortices, laminar to turbulent boundary-layer transition, and so on.

The wind tunnel static or fixed model test is based on Newton's principle of relative velocity formulated as early as 1687, the forces acting on a body immersed in a fluid flow are the same as the body moving through the fluid at rest or that of the fluid flows around the static body at the same relative speed. This change of reference poses a problem when studying ground vehicles on a road or tracks and aircraft operating close to the ground when taking off or landing. The relative speed of the vehicle to the ground influences the flow: it is usually known as the ground effect. Thus, in a wind tunnel where the vehicle is fixed, to reproduce this effect it is necessary that the floor representing the ground moves at the same speed as the

© Springer Nature Switzerland AG 2020
B. Chanetz et al., *Experimental Aerodynamics*,
Springer Tracts in Mechanical Engineering,
https://doi.org/10.1007/978-3-030-35562-3_2

air to avoid creating a parasitic boundary layer. To reproduce the ground effect, the vehicle tested (or its model) must be mounted above a rolling belt travelling at the simulated vehicle velocity (see Sect. 3.3).

The wind tunnels very practical experimental facilities which have quickly imposed themselves at the expense of alternative means based on the direct displacement of the object in the air either by:

- Horizontal: the case of the airplane tests carried out by the German company Siemens in 1901 on a train launched at 160 km/h, a technique also used by the Institut Aérotechnique (IAT) of Saint-Cyr-l'École in 1909 on a private railroad track of 1.4 km, also the wing profiles tests by Armand de Gramont (the Duc de Guiche) on his motor vehicle.
- Vertical: the guided free fall made in 1908 by Gustave Eiffel from the second floor of the eponymous tower.
- Combination of both, while taking advantage of gravity effect to move an object along a cable: method experienced in 1904 by Ferdinand Ferber in the valley of Meudon to launch his plane hanging on a cart sliding along a cable stretched over pylons. Gustave Eiffel also considered such a device (which he called aerodrome) from the first floor of his tower, before the realisation of his wind tunnel.

Rotational movement: the tested object is fixed at the end of a long rotating arm, this means making it possible to reach significantly higher tangential speeds by extending the rotating arm and the speed of rotation. Such a device was set up in 1906 at the IAT.

Except for a few applications of limited scope, the previous means were quickly abandoned because of their obvious disadvantages.

In general, a wind tunnel consists of a test section placed in a circuit where a stream of air is maintained by a fan or a compressor. In some wind tunnels, called blow down wind tunnel, the flow results from the discharge of compressed air stored in a tank. There are two types of wind tunnels.

2.1.2   The Eiffel Type or Open Circuit Wind Tunnel

In this arrangement, the air collected upstream is accelerated in a contraction consisting of a conduit of convergent section; then the air flows through the test section before being directly discharged into the atmosphere. The fluid motion is ensured by the suction effect of a fan placed near the outlet of the wind tunnel. A major innovation from Gustave Eiffel consisted of inserting between the test section and the fan located downstream, a divergent part called diffuser (Eiffel patent dated November 28, 1911). This device reduces drastically the power required for the operation of the installation. Its effectiveness stems from Bernoulli's law which states that pressure and velocity are inversely proportional. Therefore, the diffuser, by reducing the velocity, has the effect of compressing the air. Then the static pressure difference on either side of the fan can be much lower to compensate for when the fan is placed

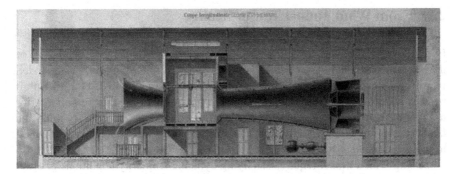

Fig. 2.1 Vintage cartoon of the Eiffel wind tunnel (© J.-M. Seguin, CSTB Eiffel Laboratory)

directly downstream of the test section. Figure 2.1 is an artistic representation of the Eiffel Wind Tunnel presently located in Auteuil, in Paris, and operated by the Centre Scientifique et Technique du Bâtiment (CSTB).

2.1.3 The Prandtl Type or Closed Circuit Wind Tunnel

In this installation, the air, after being sucked downstream of the test section and the diffuser is guided through four successive corners, and then recirculated again through the settling chamber (see Fig. 2.2). Turning vanes arranged in the corners of the circuit ensure that flow separation and the formation of vortices are avoided. This arrangement leads to improved energy efficiency and allows better control over the test conditions (pressure, temperature, moisture). This type of installation may require a heat exchanger in order to cool down the air flow which is being heated while extracting energy from the fan or compressor to maintain the circulation of the air flow.

Fig. 2.2 Installation of a closed circuit wind tunnel

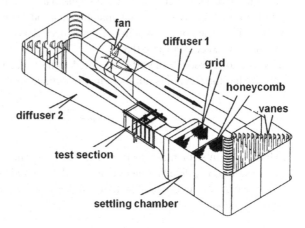

2.2 From Wind Tunnel Test to Reality

Earlier ground tests were conducted while considering real operating conditions of
the aircraft so this led to wind tunnels being designed for testing full-scale model.
However, with the increase of the flight speeds, altitude and the size of the aircraft, it
became necessary to test wind tunnel models whose dimensions can be well inferior
to those of the real vehicle. On the other hand, in the automotive industry full-scale
land vehicles are widely tested, but in these cases the test speeds are lower compared
to those encountered in aviation.

Testing with a model requires satisfying dynamic similarity conditions in order
to be able to extrapolate the results obtained in the wind tunnel to the actual flight.
The most common similarity conditions to be respected in classical aerodynamics
are often quantified by dimensionless coefficients or parameters whose definition
results from the examination of the equations of motion. These are the following:

- For similar wall boundary conditions, the geometry of the real vehicle needs to be
  reproduced or at least very representative. However, in practice, the smaller the
  scale, the more difficult or impossible it is to represent all the geometrical details of
  the real object (fasteners, panel connections, antennas, scoops, etc.). Wind tunnel
  models are thus similar only from a point of view of the global shapes, which are
  usually sufficient to determine the aerodynamic forces with the exception of the
  drag which necessitates further details of the surface features.
- At high speeds (Mach number greater than approximately 0.5); the simulation of
  compressibility effects makes it necessary to use a fluid with the same thermodynamic properties as during operating conditions. This constraint does not exist at
  very low speeds where the effect of density can be neglected and thus eliminated
  from the governing equations (except in the presence of heat transfer or reactive
  flows). In this case representative tests can be carried out in water, in hydrodynamic
  flumes or water tunnels (see Sect. 3.4), which facilitates flow visualisation.
- The reproduction of compressibility effects (in particular shock waves) imposes
  a very high accuracy on the Mach numbers between operation and wind tunnel
  test. In addition, the crossing of the sound barrier and the passage to supersonic
  flight are marked by significant changes in aerodynamic forces and their point of
  application. The compressibility effect is necessary as soon as the Mach number
  of the flow exceeds a value 0.3.
- In order to reproduce the effects related to the fluid viscosity, it is necessary to
  ensure that the Reynolds number already defined in Sect. 1.2 is matched:

$$Re = \frac{\rho V L}{\mu}$$

which accounts for the density of the gas ρ, V, the velocity, L, a characteristic
dimension of the physical model and, μ, the dynamic of the gas. Most often, the
reference quantities are relative to the uniform upstream flow of velocity V_∞, with
$\rho = \rho_\infty$ and $\mu = \mu_\infty$. For an aircraft cruising at a Mach number of 0.8, the

Reynolds number ranges between 30 and 200 millions and; for a car travelling at 120 km/h, it is approximately 9 million. Unfortunately in most of the wind tunnel test, the Reynolds numbers are usually limited to 10–100 times lower.

The Reynolds number plays a key role in representing the viscous effects, that is to say the behaviour of the boundary layers and more generally of the dissipative flow field, including wakes, mixing layers, laminar to turbulent transition, separation, etc. These phenomena can strongly influence the behaviour of the vehicle, causing a loss of performance or stability, unwanted vibrations or in worst cases more catastrophic scenarios. Except for wind tunnels where full scale vehicles can be tested, in aeronautics testing is most often performed on small-scale models, the reduction ratio being in some cases over 100. This results in a much lower Reynolds number than in flight. However, to compensate for the low Reynolds number due to the reduced dimensions of the model, it is generally not possible to increase the velocity by the necessary proportions, as it would result in a prohibitive increase in the power required. As well as the risk of entering the supersonic regime where other complications can be encountered and especially if the vehicle is supposed to operate in subsonic conditions.

Referring to the equation of state:

$$\rho_\infty = \frac{p_\infty}{r T_\infty}$$

it is possible to change the density by increasing the stagnation pressure of the wind tunnel which is then said to be pressurised (see Sect. 3.1.6). The process has limitations because the dynamic pressure of the flow,

$$q_\infty = \frac{1}{2}\rho_\infty V_\infty^2 = \frac{\gamma}{2} p_\infty M_\infty^2$$

increases by similar proportions, the magnitude of the aerodynamic forces here may introduce further problems due to deformation of the model and its support in the test section. In addition, the structure of the wind tunnel itself is then subjected to greater loading, which increases the cost of the materials of the installation and operation. Thus, pressurisation has well-defined constraints.

A very effective solution (if not simple) is to reduce the temperature of the gas which has the effect of increasing its density, as shown by the gas equation of state. This is the principle of cryogenic wind tunnels. The solution is very attractive, because while increasing the Reynolds number, reducing the temperature has no effect on the aerodynamic forces, the dynamic pressure being independent of the temperature (see relation above and Fig. 2.3). In addition, for a gas such as air, the molecular viscosity decreases with temperature: thus cooling also increases the Reynolds number.

In some wind tunnels pressurisation and cooling are combined to achieve Reynolds numbers close to those of aircraft in flight (see Sect. 4.3.4).

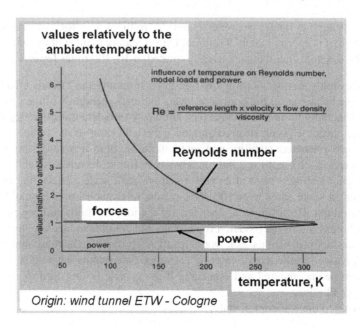

Fig. 2.3   Effect of temperature on the performance of a wind tunnel (© ETW)

2.3   Reynolds Number Effect and Laminar to Turbulent Transition

Due to the insufficient Reynolds number in most wind tunnels, the laminar to turbulent transition of the boundary layer developing on the model wall can occur further downstream compared to flight. Yet, the state of the boundary layer has a decisive influence on transfer phenomena (skin friction, wall heat transfer) as well as on boundary-layer separation and shock-wave/boundary-layer interactions. On an actual wing profile, the laminar to turbulent transition, if any, occurs very close to the leading edge (a few percent of the chord), whereas it is further downstream on the profiles tested in a wind tunnel (often at mid chord or further downstream), when tested a lower Reynolds number. One method to overcome this difficulty is to trigger transition by artificial roughness on the model, a technique often referred as boundary layer tripping. The height of the roughness or boundary layer trip devices (sometimes called turbulators) is adapted to the thickness of the boundary layer at the trip location and to the Reynolds and Mach numbers of the test.

The principle of the trip device is to fix the location of transition; its height is of similar order to the boundary layer and can be calculated using boundary-layer theory. For 2D wings, the transition can be tripped by a simple wire glued to the wall near the leading edge. A stripe of tape works in some cases and a rough band consisting of carborundum grains glued to the wall is also used. The more recent trip device consists of a thin saw tooth tape. The jagged pattern is effective at generating

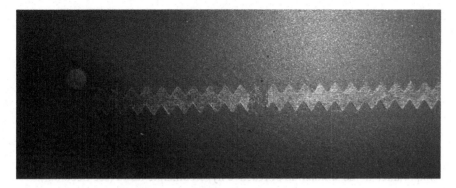

Fig. 2.4 Turbulator or transition trigger for the boundary layer

micro-vortices which destabilise the boundary layer. Figure 2.4 shows a saw-tooth transition trip placed on an aerofoil model.

2.4 Dynamics Similarity and Dimensionless Parameters

In addition to the Mach and Reynolds numbers, other dimensionless parameters must be respected during wind tunnel test, the most common being:

- **The Prandtl number** to simulate thermal conduction: $Pr = \frac{\mu C_p}{\kappa}$, C_p being the specific heat at constant pressure and κ the thermal conductivity.
- **The Lewis number**, to reproduce the diffusion phenomena of species: $Le = \frac{\kappa}{\rho D_j C_p}$, D_j being the diffusion coefficient of species j.
- **Knudsen number** which measures the ratio between the average molecular free path λ in the gas and a characteristic dimension L of the vehicle: $Kn = \frac{\lambda}{L}$. This number defines the rarefaction effects encountered at very high altitude flights, during atmospheric re-entry (see Sect. 1.2).
- For periodic or unsteady phenomena, the **Strouhal number** is introduced: $St = \frac{fV}{L}$ where f is a characteristic frequency. This number represents the ratio between the time taken by a fluid particle to travel the reference length and the period of the phenomenon.
- **The Froude number**, $Fr = \frac{V_\infty}{\sqrt{Lg}}$ whose square is a measure of the ratio between the dynamic pressure terms and the gravitational forces within the fluid. A large Froude number implies that the action of gravity is negligible as a first approximation. This number plays an important role in hydrodynamics, particularly in air-water interface problems involving a light fluid (air) and a heavy fluid (water). Since air is much lighter, most of the time the effect of gravity is neglected in problems involving air, except in modelling the Earth's atmosphere or if the weight of the displaced air is comparable to aerodynamic forces, which may be the case for airships. In general, in aerodynamics, the weight of the displaced air is ignored.

- **The Damköhler number**, $Da = \frac{KL}{V_\infty}$, where K is the reaction rate of the species considered, represents the ratio between the chemical reaction rate and the aerodynamic velocity. If this number is large, then over the distance L, a great quantity of the considered species considered could form (in the algebraic sense), or in more precise terms, the chemical reactions involving these species have had time to advance. If the reaction rates are much greater than the aerodynamic velocities, the chemical equilibrium has time to settle at any moment. On the other hand, if this number is small, within the length L, the delay, or residence Δt time, is too short for the chemical reactions to have time to advance appreciably. Then the composition of the gas does not vary: it is said frozen. This number is of great importance in the simulation of reactive hyperenthalpic flows (see Sect. 6.3).

2.5   Constraints of Testing in a Wind Tunnel

2.5.1   Effects of Blockage in the Test Section

The wind tunnel tests are performed in a test or working section, constituting of a confined space, hence a risk of the influence of the walls on the flow around the model: these are the wall effects which are very detrimental in transonic flows (see Sect. 4.2). At subsonic speeds, the size of the test section defines the longitudinal and transverse dimensions of the body to be studied. In general, the length of the test section must be at least twice the total length of the model for a diffuser length at least equal to the test section length. As far as transverse dimensions are concerned, the test section blockage effect, defined by the ratio of the largest model cross-sections area and that of the test section, should be less than 0.16. In any case, corrections to the measurements may be necessary to take into account the influence of the localised acceleration generated by the presence of the model or vehicle in the test section. Wind tunnels for land vehicles are generally equipped with a suction systems aiming at reducing the thickness of the boundary layer at the test section entrance, thus better reproducing the actual flow conditions under and around the vehicle (see Sect. 3.3).

The purpose of wind tunnel tests is most often to reproduce the flight conditions of an aircraft which operates in an unconfined atmosphere. However, in a guided test section, solid walls constrain the flow, while in an open test section the air stream is restricted to a jet developing in an external environment at constant pressure. If the dimensions of the model are not small compared to those of the test section, or the cross-section of the jet, the proximity of these solid walls or fluid boundaries respectively modifies the flow around the model, which no longer corresponds to that desired. When the size of the model is reasonably large (the limit depends on the accuracy of the desired results), interference due to walls can be evaluated and corrected to transpose to free flight conditions. For a given wind tunnel, the advantage of such an approach is to allow the test of larger models (thus higher Reynolds numbers), hence more detailed and accurate measurements.

The main wall effects that can be identified and corrected are the following:

- The presence of solid walls: the solid blockage due to the reduction of the cross sectional area or volume of the section available for the airflow and the resulting increase in speed and dynamic pressure. This effect is often negligible in open jets due to the fluid boundary conditions.
- "Wake blockage" similar phenomenon to that of solid blockage but due to displacement effect of the wake.
- Modification of the local incidence along the span of a loaded wing, mainly at the tips (opposite effects in solid or fluid boundaries).
- Modification of the natural curvature of the streamline due to localised contraction or divergence created between the model and walls, with direct effect on the overall angle of incidence, the lift and the pitching moment (weaker effects in fluid boundaries in open jets).
- Modification of the model deflection due to different aerodynamic loads on the wing with effects on the lift, drag and static stability (opposite effects in solid or fluid boundaries in open jets).
- More severe effects on propellers or rotors due to the large diameters. Significantly smaller scale models are required; other complexities are then introduced due to lower Reynolds number regime of testing (effects weaker in fluid boundaries in open jets).

One can account for a solid wall blockage by fictitiously imposing a curvature to the solid wall based on the mirror image of the model surface curvature. By reproducing the process for different configurations, it is possible to simulate the blockage effect in the wind tunnel test section (which theoretically leads to an infinite number of configurations). Using this technique the raw or uncorrected experimental results can be extrapolated to free flight conditions which takes place in the absence of wall boundaries. The influence of test section confinement is discussed in more detail in Chap. 13.

2.5.2 Model Installation and Different Kinds of Support

Another issue of parasitic interactions results from the device supporting the model that obviously does not exist in flight. This support assembly, the size of which is a function of the aerodynamic loads exerted on the model, should be as small as possible while permitting model displacements under loads. Based on the dimensions of the supports they can perturb the flow and affect the measurements on the model. In some cases, it may be necessary to carry out tests with different support configurations.

The first precaution is to make this assembly as aerodynamically discrete as possible and to fix it to the model in the least intrusive way. When these support interference effects are small, a correction of the raw measurements can be introduced by taking

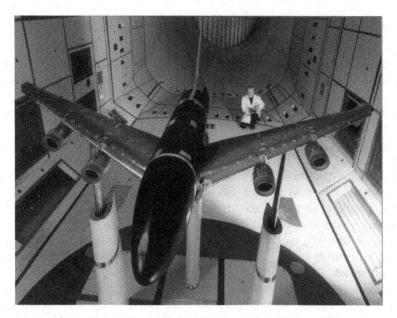

Fig. 2.5 Model of the Airbus A380 mounted on struts in the DNW wind tunnel (© DNW)

into account, through calculation or experimentally, the differences between a configuration with the support and the contribution of the support only (see Chap. 13). The main types of support or mounts used are as follows.

Strut supports: the model is supported by one or several struts often linked to a force balance placed under the test section floor. Figure 2.5 shows an Airbus A380 model mounted on struts in the Deutsch Niederländische Windkanäle (DNW) wind tunnel. The drawback of this set-up is that it perturbs the flow in the region between the model surface and the strut fixtures; this junction also promotes the formation of vortices which is a source of vibration.

Rear sting support: to avoid the above interferences, the model can be held by a sting attached to its rear part and secured to a support placed downstream (see Fig. 2.6). This avoids interferences on most of the model, except at the rear, which limits this set-up for rear-body studies. The aerodynamic forces are measured by a balance arranged inside the sting. If the model is small, miniaturised balances should be used (see Sect. 8.2). The cabling from the force sensors can be all packed inside the mount.

Nose sting support: in order to avoid disturbances of the supports or rear stings in the study of after body and base flows, the model is held by a sting fixed upstream of the test section in the wind tunnel convergent section. At large Mach number, where an axisymmetric supersonic nozzle is used, the contour of the nozzle is calculated by taking into account the presence of the central sting (see Fig. 2.7). With this type of support, it is difficult to vary the incidence of the model and it promotes an excessive

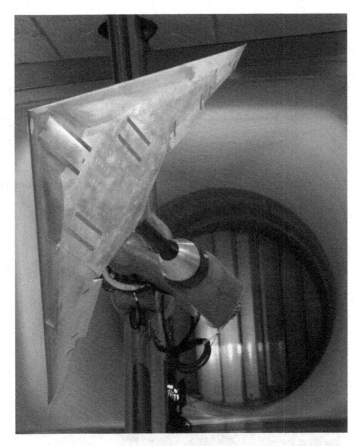

Fig. 2.6 Delta wing model mounted on a rear sting in the F1 wind tunnel at ONERA, Fauga-Mauzac
(© ONERA)

development of the boundary layer along the sting, an issue that can be addressed by
boundary-layer suction upstream of the base.

Wall mounting: for experiments on wings or aerofoils, the model is fixed to one
of the wind tunnel walls. In this case, the balance can be installed outside of the test
section. The wing (or the wing portion) is sometimes fixed to the two sides of the
wind tunnel (see Fig. 4.5). The model can also be mounted on the floor, or ceiling, of
the test section (see Fig. 2.8). If the device being tested is symmetrical, a half model
can be used, which allows doubling the size of the model for a better representation
of the details or to increase the Reynolds number.

Cable suspension: the suspension of the models by cables is an old idea, recently
reconsidered due to benefits offered by modern computer controlled systems. In the
system developed at ONERA, the suspension is a parallel redundant cable robot
(see Fig. 2.9). The model has a length of about one meter and can revolve within
a volume of $2 \times 2 \times 2$ m^3; it is held by a support beam in carbon fibre connected

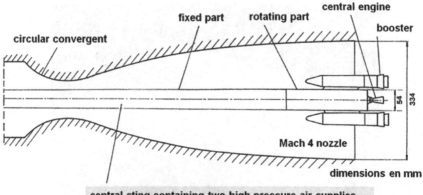

Fig. 2.7 Model of the Arianne 5 space launcher after body installed in the R2Ch wind tunnel at ONERA, Meudon (© ONERA)

to a six-component balance. The beam is connected to the wind-tunnel structure by nine motorised pulley systems to control the cables. The location of the model is determined by hybridisation of the measurements from three gyro meters and seven accelerometers embedded in the model, and also from the measurements of the lengths of the nine suspension cables and the processing of images obtained from two cameras. The determination of the aerodynamic forces exerted on the model results from a digital processing of all the measurements performed:

Fig. 2.8 Half model mounted on the floor of the F1 wind tunnel at ONERA, Fauga-Mauzac (© ONERA)

Fig. 2.9 Combat aircraft model suspended by wires in the vertical wind tunnel at ONERA, Lille (© ONERA)

- The six components balance measures the forces exerted by the support beam on the model.
- The seven measurements from the accelerometers and the three measurements from the gyro meters allow the reconstitution of the matrix of the model inertial forces.
- The displacement system gives the positions, attitudes and velocities of the model.

The wire suspension technique offers the possibility of carrying out complex tests close to free flight conditions; this new test method enables the study of the synthesis of control surface loads to identify aerodynamic forces that are usually difficult to access.

Magnetic levitation: to avoid interferences from a solid support, systems to position the model in the test section by magnetic fields were developed in the 50s. This principle based on magnetic levitation consists of placing a ferromagnetic model in an electromagnetic field generated within the working section. By varying the magnetic field intensity on each side of the test section or any direction the model can be oriented in the desired position or attitude. In the magnetic levitation system developed by ONERA, the position of the model in space is detected by light beams reflected on photodiodes. Any displacement of the model causes a current change in the cells incorporated in a feedback circuit, which induces a change in the current in the coils and thus the magnetic field intensity so as to bring the model back to its nominal position (see Fig. 2.10). The set-up developed at ONERA was very reliable, even at high Mach numbers. However, the technique had serious drawbacks, including the shape of the model that could not be too complex, the complexity of the installation and the limited measurement on the model which could be transmitted. These limitations would not have been an issue today. It is to be hoped that this ingenious solution will regain renewed interest in view of the remarkable progress made in the field of electromagnetics, electro-optics, data acquisition and processing techniques (see Chaps. 9–11).

2.5.3 Freestream Vortical and Acoustic Perturbations

Earlier issues due to low Reynolds number limitation and unwanted laminar to turbulent transition during wind tunnel testing were discussed. However, wind tunnel study of stability and transition of laminar flows is even more complicated due to freestream perturbations generated in the tunnel as opposed to the quiet environment at cruise and other flight conditions where they are assumed to be very low as the air is stationary. The perturbations in the tunnels are in the form of vortical structures which are generated upstream and/or acoustic waves which can propagate in all directions. These perturbations can either trigger the growth of instability modes in the flow through the process of receptivity or accelerate the transition process especially in the presence of high freestream turbulence. Due to these effects the wind tunnel results cannot be extrapolated to flight conditions even for similar Reynolds number

Fig. 2.10 Magnetic levitation of a model in the R2Ch wind tunnel of the ONERA Meudon centre (1965) (© ONERA)

regime as the transition modes are not influenced by the freestream turbulence, but could be affected by different noise sources from the engine for instance. This issue is more severe during transonic and supersonic tests. In an attempt to remedy these problems, "quiet" wind tunnels have been developed where special precautions are taken to attenuate the noise and disturbances caused by the installation (see Sect. 5.4).

The noise produced by the flow over a vehicle is an inconvenience for both the passengers and the surrounding environment and is therefore the major concern in the field of aeroacoustic. In order to study the sources of noise it is necessary to take special measures during the design of the test section, as well as the return circuit, as the acoustic waves emitted at the level of the model are reflected by the solid walls. The motor and fan or compressors also produces parasitic noise, often very high in amplitude. Wind tunnels intended for aerodynamic noise studies must therefore be subjected to specific acoustic treatment and are normally equipped with noise absorbing or attenuation systems (see Sect. 3.2).

2.6   The Main Sections of a Wind Tunnel

The variation of the velocity, V, of the flow in a duct of variable cross section area, A, is related by the relation given by Hugoniot's theorem based on the mass-flow conservation:

$$\frac{dV}{V}\left(1 - M^2\right) + \frac{dA}{A} = 0$$

where M is the Mach number.

In subsonic flows ($M < 1$), A and V are inversely proportional and the velocity increases in a convergent section; while in the supersonic regime ($M > 1$), the velocity increases in a divergent section. When the Mach number is equal to 1, dA and $(1 - M^2)$ must simultaneously be zero at a stationary point and the cross-section area reaches a minimum: this is normally referred as a throat (see Fig. 2.11). In a subsonic wind tunnel, the acceleration of the flow to the desired velocity takes place in a contraction section whose area decreases continuously, whereas supersonic velocities require a contraction section first to accelerate the flow to Mach 1, followed by a diverging section to continue to accelerate it to higher supersonic speed (see Sect. 5.1).

During wind tunnel test, the energy dissipated along the walls or around the model converts from pressure to heat losses by viscous dissipation effects. It is therefore necessary to maintain a pressure difference between the upstream and downstream region. In continuous wind tunnels, this is achieved by continuously running a compressor, or by discharging a reservoir of compressed air like in blow down facilities.

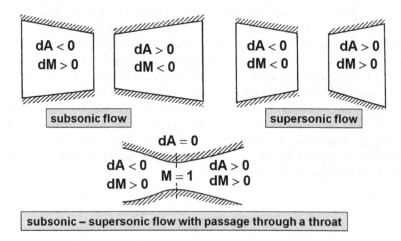

Fig. 2.11 The Hugoniot theorem and the law of section variation

The wind tunnels can be classified according to the flow velocity they achieve.

- **Subsonic wind tunnel** can operate up to 100 m/s (incompressible flow): suitable for ground vehicles, low-speed aircraft and drones, aircraft during take-off or landing phase, civil engineering applications and energy production to quote a few. There are specialised subsonic wind tunnels dedicated to study severe weather conditions such as rain, snow, sand ingestion, ice accretion and vertical wind tunnels for the study of extreme flight manoeuvres. Anechoic wind tunnels are intended for aeroacoustic studies. There are also wind tunnels dedicated to study the aerodynamics of athletes and for other sports applications.
- **Transonic wind tunnels** operate at $0.7 < \text{Mach} < 1.3$: they are mainly used for civil transport aircraft, combat aircraft, projectiles, etc. The transonic regime consists of both a subsonic and supersonic flow, which make them particularly complex. This regime is commercially sensitive because it is the usual flight regime of commercial and business aircraft which has the biggest share of the aviation industry. It also concerns propulsion systems: flow in turbomachines, engine intakes and propulsive nozzles, etc.
- **Supersonic wind tunnel**, $1.6 < \text{Mach} < 5$: nowadays are mainly used for combat aircraft, missiles, space launchers in the atmospheric flight phase and ammunitions. For commercial applications it was previously used during the design of the Concorde but renewed interest in supersonic commercial flight might lead to an increase of their usage.
- **Hypersonic wind tunnels** cover several flight regimes, $5 < \text{Mach} < 10$ for hypersonic vehicles such as those of the X-series, missiles and space launchers; $10 < \text{Mach} < 25$ highly hypersonic speed where thermal control is the predominant design consideration; $\text{Mach} > 25$, regime of operation of re-entry vehicles for which an ablative heat shield is required.

There is a clear boundary between the subsonic and supersonic regimes, characterised by the radical change of the flow behaviour when the local speed of sound is attained and this normally results in the appearance of phenomena such as expansion and compression waves, and shock waves. On the other hand, the demarcation between supersonic and hypersonic is vaguer, the Mach number being in both cases supersonic. The truly distinctive character of the hypersonic regime is the heating of the body due to the high-speed flow where the kinetic energy is transformed into heat which is more severe around the stagnation region on the body. This heating also affects the chemical behaviour of the air with appearance of so-called real gas effects resulting from the non-equilibrium between the degrees of freedom of the molecules (translational, rotational, vibrational) and chemical decomposition at very high temperature.

A subsonic or supersonic wind tunnel is made of the following parts, from upstream to downstream.

- **The settling chamber** where the freestream air is stabilised before entering the test section. This chamber is equipped with honeycomb cells to straighten the flow and turbulence screens or meshes (see Fig. 3.7) to reduce the size of the turbulence structures and the overall turbulence intensity in the test section. In a descent wind tunnel, the freestream turbulence of the flow should be below 0.1%, whereas in the ambient atmosphere where the air is at rest it is 10 times lower. As mentioned above, these higher values in the wind tunnel can be an issue for the study of the stability of flows and laminar to turbulent transition in most facilities (see Sect. 5.4).

- **The contraction section** convergent part, accelerates the flow at the entrance of the test section. In supersonic tunnels, the contraction section is followed by a convergent-divergent nozzle that accelerates the flow to the supersonic regime (see Chap. 5). The contraction ratio is defined as the ratio between the inlet and the outlet of the contraction section. This section is an important component of the wind tunnel and the wall curvature must be carefully design to ensure a good homogeneity of velocity throughout the test section. It also helps in reducing the size of the freestream turbulent scale entering the test section. A contraction ratio close to 10 is recommended for a tunnel with a decent flow quality.

- **The test section** is where the model is mounted and as we know there are two types: the guided test section with walls and the open test section constituting of a free jet emerging from the contraction section and then followed by the diffuser. The open test section has the advantage of providing direct access to the model, which facilitates observation of the flow field and measurements by optical methods. Its disadvantage is the shear layer instability and the turbulent mixing at the boundary of the jet which is a strong source of unsteadiness and acoustic disturbances. The guided test section does not allow direct access to the model, this can hinder observation of the flow around and on the model (for visualisations and optical measurements). It has the advantage of a better flow quality due to the absence of the shear layer. The section of the guided test section is often slightly divergent so as to maintain a constant velocity by compensating for the thickening of the boundary layers on the walls. The initial tendency was to use a circular shape as test section, for structural reasons and better quality of the flow by avoiding corner vortices. The curved wall, on the other hand, has the disadvantage of complicating optical measurements. For this reason, and also to facilitate installation of the models, rectangular test sections tend to impose themselves.

- **The diffuser** consists of a divergent duct widening towards the fan or the exhaust in the open circuit tunnel. To reduce the size of the installation, it is advantageous to adopt a diffuser which is as short as possible, so a high angle of divergence. However, too great a divergence causes a rapid increase in pressure; therefore a risk of separation of the boundary layer on the diffuser wall, resulting in a significant efficiency loss and noise. The design rule that was imposed very early is not to exceed a divergence angle of 6°–7°.

- **The motor and fan unit** restores the pressure level of the flow before ejecting it into the atmosphere (Eiffel type tunnels) or reintroducing it into the return circuit (closed loop tunnels). For high Mach number wind tunnels, which require a very high pressure ratio, the air flow is often produced by the discharge of air stored at high pressure. These installations are called blow down facilities, the useful test time being limited in particular by the storage capacity.

In general, wind tunnels can be classified into two categories:

- **Industrial wind tunnels** often large in size, are intended for testing more representative configurations such as the model of a complete aircraft or even real full scale vehicles in automotive applications.
- **Research wind tunnels** are of small or medium size, devoted to fundamental aerodynamic studies with the aim of characterising particular complex phenomena such as laminar to turbulent transition, separation, shock-wave/boundary-layer interaction, turbulence properties, etc. or to evaluate new concepts, for example in the field of flow control.

However, the distinction is sometimes arbitrary as fundamental tests can be also performed in large wind tunnels to achieve realistic Reynolds numbers or work on a larger model better suited for the measurement of fine flow structures or quantities. Conversely, applied studies are often performed in research wind tunnels because of their better flow quality and cost of running.

2.7   Design and Manufacturing of Wind Tunnel Models

The large wind tunnels built until the middle of the 20th century, such as S1MA at ONERA, Modane-Avrieux (see Sect. 4.3.1) were intended to test full scale aircraft while avoiding problems related to Reynolds number in particular. The progress in numerical simulation to handle some of the issues related to wind tunnel testing of larger full scale models to satisfy the demand of the market and the cost of running these facilities increased rapidly. This led to serious reduction usage of these facilities and several of them being made redundant. However, the automotive sector did not suffer similar trend as there are more modern wind tunnels for testing full scale vehicles even under more severe climatic conditions (see Sects. 3.2 and 3.3). But for the aerospace industry, the use of scaled down models is the general rule, with all the complexity due to dynamic similarity covered in this book. The design and construction of a wind tunnel model is therefore a task with very high added value on which the quality of the results depends to a large extent.

Geometric representativeness is the major requirement. It is still necessary to choose the adequate level of representation since all the details of a real plane cannot be represented and would not be, given the Reynolds number effects. Conversely, additional features might be required to trigger boundary-layer transition. It is also necessary to take into account the deformation of the model during the tests. The degree of roughness of the aerodynamic surfaces must be also well controlled during the manufacturing process which sometimes conditions machining technology or might require manual finishing.

As weight is generally not a problem in the wind tunnel, models are often made of solid steel to minimise deformation (see Fig. 2.12). Still, some deformations can be significant and it is necessary to identify them during the tests (see Sect. 9.6). On the other hand, to limit the weight of the models and the cost of realisation,

Fig. 2.12 High Speed Machining (HSM) of a fuselage model (© ONERA)

aluminium is used for fuselages or parts which are not subjected to high loading. Moreover the interior of the model cannot be totally filled so as to accommodate instrumentations (see Fig. 2.13). A good example is the aerodynamic balance (see Chap. 8) to measure the forces and the moments acting on the model. The necessary stiffness of the balance limits its miniaturisation and it its dimensions might constrain the size of the model.

For static pressure measurement on an aircraft model it may require more than 1000 pressure ports (tapings or tubes with a diameter of less than 1 mm), particularly on the wing. These tapping are connected individually to the pressure transducers by metal and/or plastic tubes while passing inside the model together with additional electric wires for various other sensors (balance, inclinometer, pressure transducers steady and unsteady, etc.).

Hence a considerable volume of cables and pipes has to be taken into account in the internal layout of the model (see Fig. 2.14). This resulted in instrumentation complexity which led to the development of transducers with built-in pressure tap and signal conditioning so that the output is digitised and acquired directly; this greatly reduces the space occupied by the electric wiring. The disadvantage is the high cost of design and manufacturing of the model, therefore this type of transducer are reserved for large models which are highly instrumented.

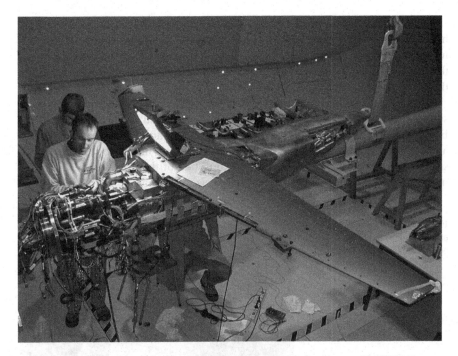

Fig. 2.13 End of integration of the instrumentation in a model before the test (© ONERA)

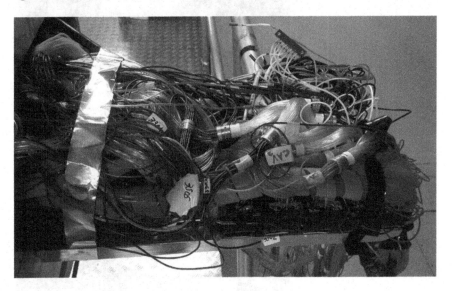

Fig. 2.14 Tubing inside a model equipped for pressure measurements

Today, machining accuracies of the order of a few microns can be achieved using high-speed steels, but these processes are expensive. Additive manufacturing techniques should open up new design possibilities and reduce costs and cycle times, but the structural integrity remains a concern and the use of more robust material drives the cost.

Wind tunnel tests are often used to verify the effectiveness of the aircraft's control surfaces and moving parts such as leading edge slats and trailing edge flaps (see Fig. 2.15). The most common practice is to set these movable parts to the desired angle of incidence to be tested. Such an operation requires intervention on the model, so the wind tunnel has to be stopped. Motorisation of these moveable surfaces is however possible on certain models of large size: in this case, the model is more

Fig. 2.15 Large size model equipped with moving control surfaces (© ONERA)

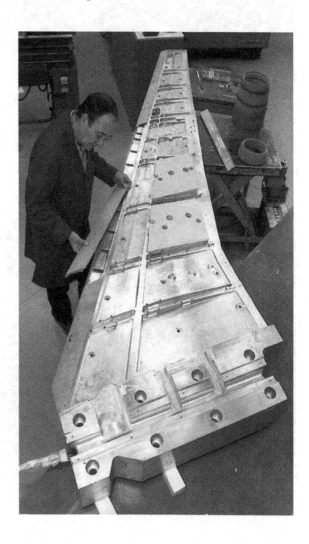

Fig. 2.16 Model for cryogenic wind tunnel equipped with motorised control surfaces

expensive but the test productivity is much improved. This is a compromise to be considered when planning designing the model.

In a cryogenic wind tunnel, like ETW (see Sect. 4.3.4), it is not possible to intervene immediately on the model exposed to a very low temperature. Therefore productivity can be increased by motorisation of certain moving parts. This requires very high precision mechanisms and actuators of very small size, capable of operating at very low temperatures (see Fig. 2.16).

In experiments on models where the influence of the suction from engine is considerable, as it modifies significantly the mean flow and the streamline patterns, a simple representation of just a dummy nacelle might not be sufficient and therefore the Turbine Power Simulator (TPS) technology has been developed. It consists of a small turbine installed in the model nacelle simulating the engine flow (see Fig. 2.17). This turbine is powered by compressed air fed through the model support.

Wind tunnel models are very high-tech hardwares thus their design and construction are an important part of the cost and cycle of experimental aerodynamics. They are usually designed based on the specification of the wind tunnel in which they will be tested, this determines the size of the model and the materials if the wind tunnel is pressurised and/or cryogenic. But also on the objectives of the test which determine whether the parts will have to be interchangeable, the need for motorisation, the type of instrumentation to be installed in the model and how it will be mechanically supported during the test. Since the complexity is not limited to the aforementioned objectives the dialogue between the team in charge of the design and the wind tunnel test engineer must therefore start very early in the design cycle of complex model (Fig. 2.18).

Fig. 2.17 Models of nacelles equipped with TPS (© ONERA)

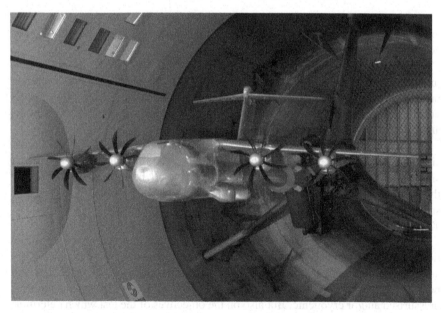

Fig. 2.18 Model of the A400M military transport aircraft in the S1MA wind tunnel of the ONERA Modane-Avrieux centre (© ONERA)

Chapter 3
Subsonic Wind Tunnels

3.1 Subsonic Wind Tunnels for Aeronautics

3.1.1 Historical Subsonic Wind Tunnels in France

Built by Gustave Eiffel, and still in use, the low speed wind tunnel at the Scientific and Technical Centre for Building (CSTB) located in Auteuil was inaugurated on 19 March 1912 (see Fig. 2.1). Figure 3.1 shows Gustave Eiffel and his colleague Léon Rith in the wind tunnel control room.

This installation is a transfer of the first wind tunnel, installed in 1909 at the foot of the Eiffel Tower, in the Champs-de-Mars. Initially it achieved a velocity of 18 m/s in a circular test section of a diameter of 1.5 m and was driven with a fan powered by a 37 kW motor. With the same motor, during the re-installation at Auteuil the new facility was equipped with a diffuser and allowed the tunnel to reach 32 m/s (an increase of 78% of the velocity) in a test section of a diameter of 2 m, resulting in a mass flow gain of over 200% (see Fig. 3.2). This marked the birth of the diffuser which was a major improvement in wind tunnel design and helps to increase the speed at which they could be operated by offering a recovery in pressure.

The wind tunnel is still used for industrial aerodynamics studies, wind effect on buildings, aeroacoustic and other aerodynamic applications as shown in Fig. 3.3.

Although no longer in service, as an introductory remark it is interesting to briefly describe this typical very large facility inaugurated in 1935 on the site of Chalais-Meudon and originally designed for full scale aircraft testing. Figure 3.4 shows an aerial view of the wind tunnel. Of the Eiffel type, this wind tunnel has a contraction ratio of 3.5 with an inlet of 100 m^2 at the beginning of the working section. The collector captures the air from outside the facility through a large plenum which is then accelerated to a maximum speed of 45 m/s. The exit of the open jet test section has an elliptical shape with a 16 m major axis and an 8 m minor axis. Its diffuser consists of a truncated cone-shaped tube with a length of 38 m, made of 70 mm thick reinforced concrete. Downstream of the diffuser, the suction chamber is equipped

© Springer Nature Switzerland AG 2020
B. Chanetz et al., *Experimental Aerodynamics*,
Springer Tracts in Mechanical Engineering,
https://doi.org/10.1007/978-3-030-35562-3_3

Fig. 3.1 Gustave Eiffel in the Auteuil wind tunnel

with six 74 kW fans. The facility was decommissioned in 1977 and was listed as a historic monument in 2000. This facility allowed testing the aerodynamics of aircraft, cars, trains and buildings in the framework of many national programs.

Fig. 3.2 The CSTB Eiffel wind tunnel in Auteuil. Wind tunnel convergent inlet equipped with a honeycomb (© CSTB)

Fig. 3.3 Automotive aerodynamic tests in the Eiffel Wind Tunnel (© CSTB)

3.1.2 Circular to Octagonal Cross Section Wind Tunnel

An example of such a wind tunnel is the **S2L of the ONERA Meudon centre** which is a low speed wind tunnel of the Eiffel type driven by a 32 kW motor. Its velocity range is between 4 and 45 m/s and it has three interchangeable test sections: two of circular section and an octagonal section for optical access. The collector has an inlet

Fig. 3.4 The Grande Soufflerie of Chalais-Meudon (© ONERA)

diameter of 3.14 m and a length of 3.5 m. The circular test section has an entrance
diameter of 0.97 m and a length of 1.32 m, the total length of the wind tunnel being
20 m (see Fig. 3.5). The "optical" test section, of octagonal cross section, is equipped
with planar glass windows so as to facilitate measurements by optical techniques (see
Fig. 3.6). A circular test section has the advantage of mitigating the formation of high
intensity corner vortices from rectangular or square sections. The effects from corner

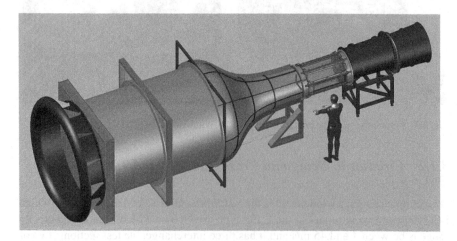

Fig. 3.5 The S2L research wind tunnel of the ONERA Meudon centre (© ONERA)

Fig. 3.6 The S2L wind tunnel equipped with an octagonal test section for optical measurements (© ONERA)

vortices can be reduced by corner fillets which eventually turn the square rectangular section into an octagonal section.

The octagonal optical-access windows allow for direct flow measurement techniques such as PSP (Pressure Sensitive Paint), LDV (Laser Doppler Velocimetry), PIV (Particle Image Velocimetry) which would necessitate corrections for spherical aberration with a curved window in a circular cross section configuration. The wind tunnel is dedicated for research for both aeronautic and aerospace applications, including studies of vortical flows, wake vortex, separated flows, blade-tip clearance, CROR (Counter Rotating Open Rotor) blades.

3.1.3 Low Reynolds Number Wind Tunnel

The **ISAE-SUPAERO Low Reynolds number (SaBRe)** facility is a closed circuit wind tunnel designed for the study of micro or macro unmanned aerial systems (UAS or MAS) in both powered and unpowered configurations to study particular issues due to propeller/wing interaction, flow control, drag reduction, etc. It is also used for studying other types of aircraft or land vehicles. It has a rectangular test section of 1.2×0.8 m^2 and 2.40 m long (see Fig. 3.7). The wind tunnel is driven by a fan

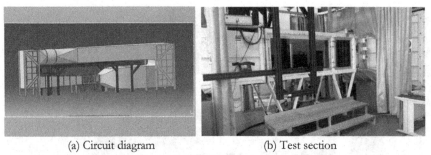

(a) Circuit diagram   (b) Test section

Fig. 3.7 The low Reynolds SaBRe Wind Tunnel (© ISAE-SUPAERO)

of 9 variables pitch blades, powered by a DC motor of 20 kW. The variable pitch fan offers the possibility of continuously generating a flow from quasi stationary to 25 m/s under stable and fairly quiet conditions since the shaft speed can be kept constant and at low revolutions.

The settling chamber is equipped with a honeycomb and 3 turbulence screens by, in conjunction to the large convergent contraction, with a ratio of 9, a high flow quality of a turbulence intensity below 0.03% is generated over the entire velocity range of 2–25 m/s. The test section side walls are made of opaque or transparent removable panels, depending on the kind of tests and the instrumentation used.

The SaBRe wind tunnel is equipped to characterise:

– the flow properties using optical measurement systems such as 2 and 3 components PIV, TR-PIV, LDV, hot wire anemometry, steady or unsteady pressures, temperature, etc.
– the overall aerodynamic forces on the model using force balances adapted to the model or support used.

It is also equipped with a robot arm allowing automated 3D displacements of the model during dynamic testing to simulate simultaneous pitch, roll and yaw motions (see Fig. 3.8).

Fig. 3.8 MAVs mounted in the SaBRe wind tunnel (© ISAE-SUPAERO)

3.1.4 Multiple Test Section Wind Tunnel

The **Malavard wind tunnel at the PRISME laboratory** in Orléans is a facility with multiple test sections, inaugurated in 1990. This facility has the particularity of having two test sections: the main test section and a secondary test section in the wind tunnel return circuit (see Fig. 3.9).

The main test section has a length of 5 m and a cross section of 2×2 m². It provides a flow velocity of 55 m/s with an adjustable turbulence intensity level of at least 0.4% (see Fig. 3.10). The tunnel is equipped with a 6-component floor mounted force balance and its transparent walls allow the use of optical diagnostic techniques.

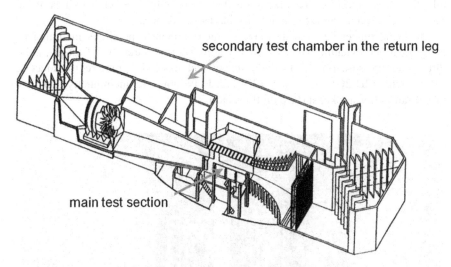

Fig. 3.9 The Malavard wind tunnel circuit of the PRISME Laboratory (© PRISME Laboratory)

(a) General view of the test section   (b) Test on a future airplane model

Fig. 3.10 Main test section of the Malavard wind tunnel at the PRISME laboratory (© PRISME Laboratory)

Specific set ups are available to realise a wide range of research and industrial tests on simple generic configurations, vehicle models (aircraft, land vehicles, ships, etc.), wing profiles and other devices.

The return circuit constitutes a secondary test section of adaptable cross section ranging between 2.5 × 2.5 m² and 4 × 4 m² with a maximum velocity of 35 m/s and an adjustable turbulence level. This test section allows testing on various types of large object and is equipped with a rotor test-bed. Thanks to turbulence generators and the development of a thick turbulent boundary layer on a 16 m long floor, this test section makes it possible to reproduce on a reduced scale an atmospheric boundary-layer flow of up to 15 m/s, for tests on building models, urban topography models, relief topography, wind turbine farms models and many other tests requiring such inlet flow (see Fig. 3.11). This facility is dedicated to the research activities of the PRISME Laboratory and is also used for industrial services.

The wind tunnel has a wide range of measurement capabilities including 6-component aerodynamic balances, rotor testing rig (measurement of torque, axial thrust and pressure on the blades), steady and unsteady pressure transducers, PIV systems (2D-2C), stereo-PIV (2D-3C), LDV, hot-wire anemometry and flow visualisation (by smoke, surface oil flow, etc.).

Fig. 3.11 Urban topography test in the return circuit of the Malavard wind tunnel of the PRISME laboratory (© PRISME Laboratory)

3.1.5 Low Turbulence Research Wind Tunnel

This type of wind tunnel is well suited for fundamental studies of complex flows and their stability such as two and three-dimensional boundary layers, separation, vortices, wakes, laminar to turbulent transition, as well as studies on flow control for maintenance of laminarity and suppression of separation.

The **F2 closed loop wind tunnel of the ONERA Fauga-Mauzac centre** is such a research facility with a rectangular test section 1.4 m wide, 1.8 m high and 5 m long (see Fig. 3.12). It is equipped with a 12-bladed fan driven by a 680 kW DC motor. The velocity can be continuously varied from 0 to 100 m/s by adjusting the motor speed. The settling chamber is equipped with four screens, and a honeycomb filter and absorbent walls which, in conjunction with a contraction ratio of 12, provide a very low level of turbulence in the test section (less than 0.05%). The test section side walls consist of opaque or transparent removable panels which can be adapted to the tests.

The wind tunnel is equipped with a laser velocimetry system allowing the simultaneous measurement of the three velocity components. Mechanical and optical variants make it possible to constitute measuring volumes according to the optical access to the measurement point (see Fig. 3.13).

The point wise measurement is fully automated and can cover a volume of $0.5 \times 0.6 \times 1.0$ m^3. Figure 3.14 illustrates a typical test in the F2 wind tunnel.

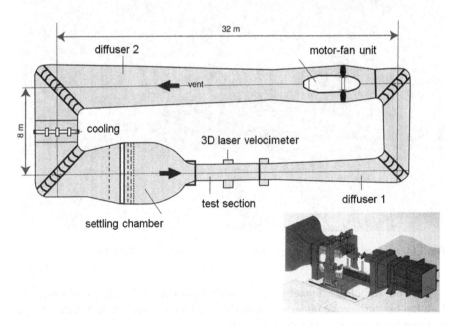

Fig. 3.12 General arrangement of the F2 wind tunnel at ONERA, Fauga-Mauzac (© ONERA)

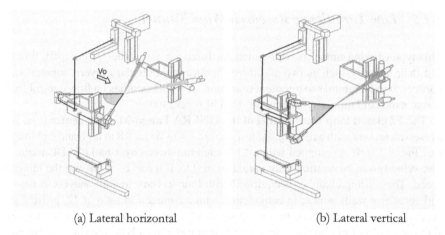

(a) Lateral horizontal   (b) Lateral vertical

Fig. 3.13 Configurations of the LDV system installed on the F2 wind tunnel (© ONERA)

Fig. 3.14 Study of vortices on a delta wing in the F2 wind tunnel (© ONERA)

The low turbulence tunnel in the **Gaster Laboratory at City, University of London** has turbulence intensity below 0.01%, but the speed range and the size of working is almost twice less than that of the F2.

3.1.6 Pressurised Subsonic Wind Tunnel

Operational since 1974, the **F1 wind tunnel** at ONERA, Fauga-Mauzac is a closed circuit wind tunnel with a rectangular test section of a cross-section of 4.5 m by 3.5 and 11 m long (see Fig. 3.15). Pressurisation of the installation (up to 3.85 bar) increases the density of the flow in order to maintain a high Reynolds number to compensate for the reduced size of the model compared to the full scale vehicle.

The F1 power-plant is a 16-blade, constants-speed fan driven by a 9.5 MW electric motor. Mach number is adjusted by changing the pitch angle of the blades. The flow maximum velocity is 123 m/s (Mach 0.36) at a stagnation pressure of 1 bar, this maximum value decreasing when the stagnation pressure is increased (see Fig. 3.16). It is not possible to achieve a factor greater than 4 on the Reynolds number by pressurisation, the dynamic forces on the structures (walls, floors) becoming then too significant. In addition, since the aerodynamic forces being also increased by the same proportions, excessive deformation of the model and a possible failure of its support could be experienced.

The test section and its cylindrical casing are mounted on carts detachable and movable allowing for rapid changeover and testing of configurations already mounted and prepared in three other interchangeable sections equipped with a data acquisition unit.

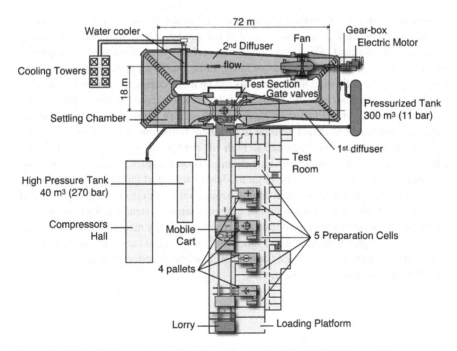

Fig. 3.15 General arrangement of the F1 wind tunnel of the ONERA Fauga-Mauzac centre (© ONERA)

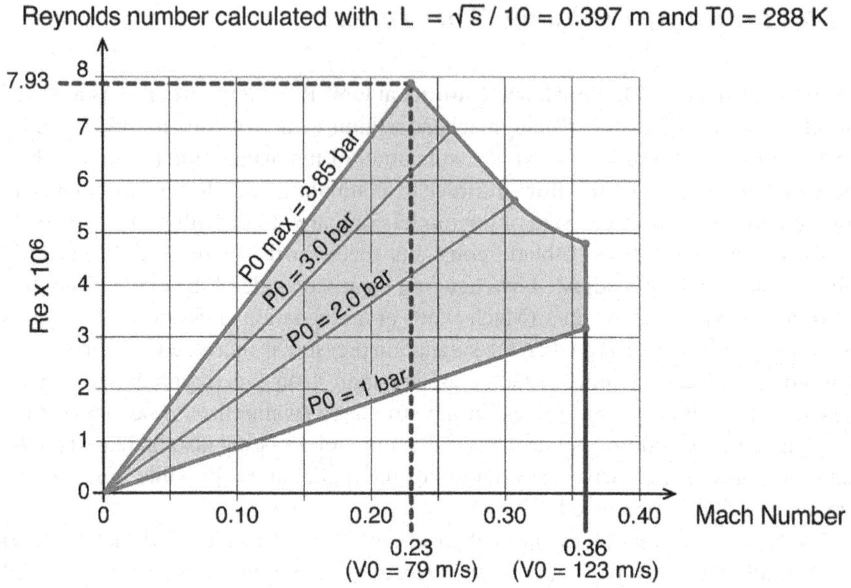
Reynolds number calculated with : L = √S / 10 = 0.397 m and T0 = 288 K
7.93
Re x 10⁶
PO max = 3.85 bar
PO = 3.0 bar
PO = 2.0 bar
PO = 1 bar
Mach Number
0.23
(V0 = 79 m/s)
0.36
(V0 = 123 m/s)

Fig. 3.16 Test range of the F1 wind tunnel (© ONERA)

Airtight sealed doors make it possible to isolate the test section from the circuit and to intervene during the changeover of the model, while the remaining circuit remaining pressurised. The wind tunnel has several types of support according to the requested tests: different kinds of sting, three masts support, wall turrets, wall balance, etc. The following measurements can be made during a continuous variation of test parameters: steady and unsteady pressures, forces (6 components), model deformation, flow visualisation by laser light sheet, coloured oil flow visualisation, sublimation product to detect laminar to turbulent transition, infrared camera, PIV.

The wind tunnel is well adapted to high-lift configuration studies, aerodynamic force and moment (6 components) measurements on complete or half-models, control surface force measurements, large models of transport aircraft (more than 3 m span) or military aircraft, tests with engine simulation (turbofan or turboprop), ground effect tests, air intake tests for fighter or transport aircraft (steady and unsteady measurements) and other specialised tests for dynamic stability measurements, building, ship, car, helicopter aerodynamics. Figure 3.17 shows a test on model support effect, note in the background the closed door isolating the test section from the pressurised circuit during installation of the model.

Fig. 3.17 Study of the model support effect in the F1 wind tunnel (© ONERA)

3.1.7 Large Research Wind Tunnels

The **L2 wind tunnel of ONERA Lille centre** is located in an industrial hall of 725 m² forming the return circuit. The test section, guided over a length of 13 m, has a rectangular section, 6 m wide and 2.4 m high. The settling chamber is equipped with a honeycomb which was added in 2015. A diffuser leads to motor-fans equipped with individual diffusers. The displacement of the probes is achieved by a sufficiently precise 3 axes motorised gantry suspended from the ceiling. On the floor, a 6 m diameter turntable helps in positioning the models accordingly to a desired angle with respect to the wind direction. The flow in the test section is generated by a set of 18 motor-fans mounted in 3 rows over a height of 2.8 m, with a total power of 125 kW (see Fig. 3.18). A maximum velocity of 19 m/s can be achieved in the empty test section.

The L2 wind tunnel is well suited to study the effects of wind on land or on the sea, thanks to the ability to simulate the atmospheric boundary layer (average speed profile). The air flow development over aircraft carriers has been studied in this wind tunnel in order to characterise and mitigate various phenomena such as turbulent levels in the landing zones, soot fallout, studies of smoke plumes for pilot visibility

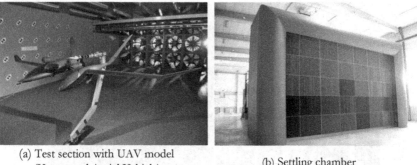

(a) Test section with UAV model
(Unmanned Aerial Vehicle)

(b) Settling chamber

Fig. 3.18 The L2 wind tunnel at ONERA, Lille (© ONERA)

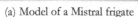

(a) Model of a Mistral frigate   (b) Laser sheet visualisation on a Mistral frigate

Fig. 3.19 Study of the effect of wind on aircraft carriers in L2 wind tunnel (© ONERA)

and interaction with electronic equipment, positioning of anemometers, protection of inhabited areas, etc. Figure 3.19 illustrates a test on a model of a Mistral frigate.

The **S1 wind tunnel of the Institute of Fluid Mechanics of Toulouse (IMFT)** was built in 1936. It is an Eiffel type installation, with a test section diameter of 2.40 m and a length of 1.80 m (see Fig. 3.20). It can reach a speed of 37 m/s by means of a 4.20 m diameter fan. This wind tunnel which was initially working in the open air was moved in a building in 1942 to protect it from climatic hazards and stabilise the operating conditions. The wind tunnel is equipped with a computerised aerodynamic balance and optical measuring systems, such as three components LDV and PIV. Since the 1980s, the S1 wind tunnel has been contributing to research projects in aerodynamics characterisation of turbulence around wing profiles, wind turbines, flow control. It is also used to study air pollution with equipment of modest size and due to its reduced operating cost it is suitable for short-distance plume diffusion studies, taking into account local effects (relief, obstacle, etc.). The ease of access in the open test section makes S1 a convenient means for local exploration of free flow fields (separated flows, simple or coaxial jets, free or sheared by a side wind, etc.). It

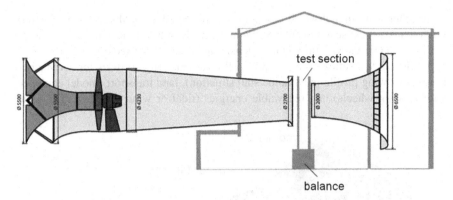

Fig. 3.20 The S1 wind tunnel of the Institute of Fluid Mechanics in Toulouse (dimensions in mm) (© IMFT)

contributes to the studies on separation control on generic automobile configurations and recently an experiment on electro-active wing morphing of future airplane was conducted.

The **S4 wind tunnel of IMFT** closed loop facility was built in the mid-1970s in order to satisfy the experimental needs and to provide complementary aspects of a closed circuit facility (low level of turbulence, air filtration). The S4 wind tunnel has a rectangular test section (0.60×0.70 m^2) with axial expansion to compensate for the development of boundary layers on the tunnel walls (see Fig. 3.21). The temperature stabilised flow can reach a velocity of 55 m/s for a turbulence level of the order of 0.1%. The S4 wind tunnel is used for fundamental studies on turbulence modelling, flow control and fluid-structure interactions. It is equipped with traverse system for hot wire probes with four degrees of freedom (three in translation and one in rotation) and advanced means of laser metrology (TR-PIV-3C and TR-PIV at high speed). The S4 wind tunnel hosts innovative prototype of aircraft wings for studies on electro-active morphing.

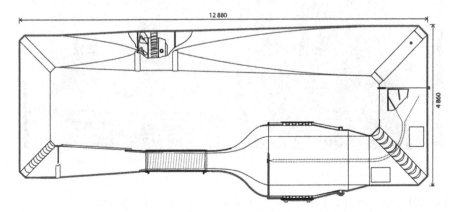

Fig. 3.21 The S4 wind tunnel of the Institute of Fluid Mechanics in Toulouse (© IMFT)

Another wind tunnel worth mentioning is the **S620** of ISAE-ENSMA. The test section has a cross section of $2.6 \times 2.4 \text{ m}^2$ and a length of 6 m. The velocity range is from 5 to 60 m/s with turbulence intensity less than 0.5% (see Fig. 3.22).

Dedicated test setups (see Fig. 3.23) allow a wide range of tests for aeronautics (aircraft or wing profile in dynamic stall situation), land transport (model equipped with rotating wheels) and renewable energies (tidal or wind turbines). The wind

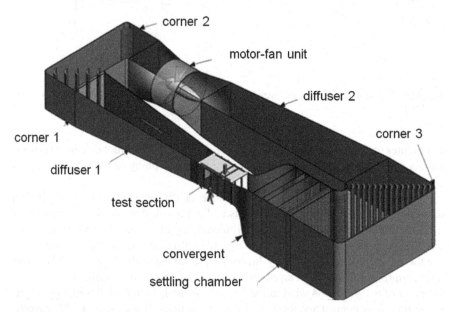

Fig. 3.22 Circuit of the S620 wind tunnel of the PPRIME Institute in Poitiers (© PRISME Laboratory)

(a) Two-dimensional set up for profile tests

(b) Floor mounting for testing of land vehicles

Fig. 3.23 The S620 wind tunnel of the PPRIME Institute (© PPRIME Institute)

tunnel is equipped with 6-component aerodynamic balances allowing the measurement of average or unsteady forces via the use of piezoelectric dynamometers (see Sect. 8.2.3). Pressure measurements on the model can be performed with steady (128 channels) or unsteady (64 channels) pressure transducers. Velocity measurements are performed by 2 or 3-component PIV (at low and high speed), LDV as well as by hot wire or film anemometry (constant temperature mode).

3.2 Special Purpose Wind Tunnels

3.2.1 Vertical Wind Tunnel

The **vertical wind tunnel of the ONERA Lille centre** is a low-speed Eiffel type wind tunnel equipped with an open test section. This installation makes it possible to generate an upward flow up to 50 m/s in a test section of 4 m in diameter. As shown in Fig. 3.24, the wind tunnel is contained in a cylinder of 11 m in diameter, with the engine room above it. A second cylinder of 17.5 m in diameter, forming the tower visible from the outside of the building, has a height of 36 m. The 800 kW

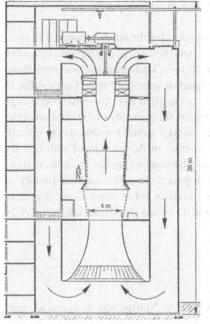

(a) Arrangement of the wind tunnel   (b) Exterior view

Fig. 3.24 The vertical wind tunnel at ONERA, Lille (© ONERA)

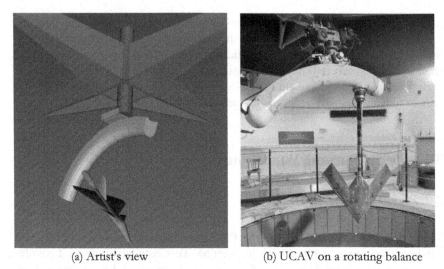

(a) Artist's view   (b) UCAV on a rotating balance

Fig. 3.25 Rotating balance of the vertical wind tunnel at ONERA, Lille (© ONERA)

motor installed at the top of the tower drives a thirteen-blade fan by a vertical shaft assembled to the major horizontal shaft of the motor. Commissioned in 1966, the wind tunnel has been largely modernised, while the original honeycomb and contraction section has been preserved.

Originally the vertical wind tunnel was intended for the study of aircraft in spin, the models being launched into a spin in the open test section were free to evolve under the influence of gravity and aerodynamic forces. This technique was abandoned for a rotating balance (see Fig. 3.25). This dynamic assembly makes it possible to reproduce a large range of attitudes of the aircraft relative its to forward speed (large incidence and yaw) and to set the model in rotation for the study of the rotation effect on the aerodynamic coefficients. This facility is well suited for forecasting aircraft behaviour at the boundaries of the flight envelope, in the vicinity of stall and for spin studies. The vertical wind tunnel can also be used to predict the behaviour of free-falling machines (probes, balloons, small parachutes), the manoeuvrability of submarines, the training of free-flying parachutists. A robot allows a six degree of freedom displacement of the models, or to simulate a semi-free flight (see Sect. 2.5.2).

3.2.2 Climatic Wind Tunnels

The **Jules Verne climatic wind tunnel of CSTB in Nantes** was designed to study full scale models the combined effects of wind and other climatic parameters such as rain, sand, sun, temperature and snow on constructions, vehicles or any other system subjected to extreme weather conditions. This installation consists of two

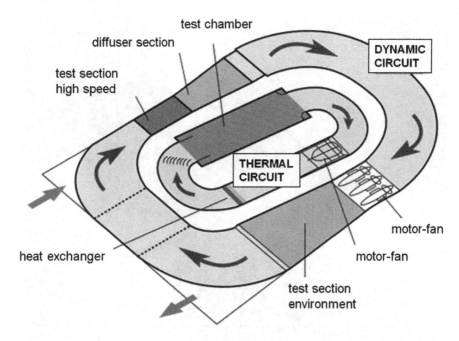

Fig. 3.26 General arrangement of the Jules Verne wind tunnel (© CSTB)

concentric rings (see Fig. 3.26). The dynamic circuit occupies the outer ring and is operated by six fans of a total power of 3 MW; a speed of 100 km/h is reached in the environment 100 m^2 test section and 280 km/h in the 30 m^2 high speed test section. The dynamic circuit can simulate large-scale turbulence structures occurring naturally in the atmosphere, rainfall up to 200 mm/h, a sandstorm due to the possibility of operating as an open section, or a tropical storm.

The thermal circuit (inner ring), independent of the outer circuit, simulates a wide range of climatic environments namely fog, ice, snow, sunshine and many types of rainfall, with an adjustable temperature of −32 to +55 °C. The maximum speed reached is 140 km/h using a 1.1 MW fan (see Fig. 3.27).

This semi-guided test section has a nozzle of adjustable section area ranging between 18 and 30 m^2. The test area has a length of 25 m, width of 10 m and height of 7 m. Heat exchangers with a cooling capacity of 2 MW controls the temperature at a rate of 15 °C/hour. This wind tunnel is equipped with a rolling road to test vehicles with realistic engine loads (see Fig. 3.28). It is one of the few in Europe to be able to reproduce the driving conditions of vehicles in snowy conditions (fine and dry snow or heavy snow).

Fig. 3.27 Fan of the thermal circuit of the Jules Verne wind tunnel (© CSTB)

3.2.3 Icing Wind Tunnel

The ice accretion is through the crystallisation of super-cooled water droplets present in the clouds, upon impact on the surface of the aircraft, in particular on the leading edges of the wing or the nacelle it can then deposit as a layer of ice altering the geometry of the wing whose lift and drag can be affected. In addition, the ice deposit can modify the weight of the aircraft which can lead to severe consequences. It is therefore essential to understand the mechanism of ice accretion to be able to design and test devices for protection or de-icing, hence the icing wind tunnel are of paramount importance.

Icing tunnels are useful to both the aircraft manufacturer and certification authorities because of their ability to duplicate certification requirements in a repeatable and controlled environment. They represent an important step in the development and certification of any anti-icing or de-icing system. In addition, they are a very useful tool for validating computer codes specialised in the simulation of ice accretion. The Icing Wind Tunnel or IWT, the largest and most advanced installation in the

Fig. 3.28 Simulation of heavy snow conditions in the Jules Verne climatic wind tunnel (© CSTB)

world, is located at the Centro Italiano Ricerche Aerospaziali (CIRA) in Capua. Its test sections can accommodate a wide range of full-scale objects: engine inlets, wing section, landing gear, weapon systems, etc. State-of-the-art instrumentation makes it possible to evaluate the effectiveness of the de-icing systems subject to certification, as well as to highlight the consequences of any malfunction that is detrimental to flight safety.

The IWT was designed to replicate the cloud conditions specified in the icing certification rules. In this installation, shown in Fig. 3.29, the test section is inter-

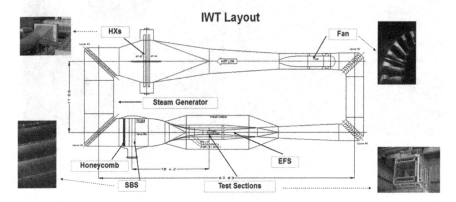

Fig. 3.29 Layout on the CIRA Icing Wind Tunnel (IWT) in Capua (© CIRA)

changeable with three different test sections (all of the same height 2.35 m and width 1.15 m, 2.25 m or 3.6 m) in order to best adapt to model size, speed, cloud cover and flow uniformity. Each configuration can be used for both aerodynamic and icing experiments by interchanging the turbulence filter module and the module generating the icing cloud.

The IWT test section, in particular the distance between the cloud generation section (spray boom) and the model stagnation region where accretion occurs, was dimensioned in order to realise the thermal equilibrium rather than the quality of the flow. The fan, including its motor, is located in the return leg of the circuit. The diameter of the fan is 3.9 m, its maximum rotation speed being 750 rpm driven by a 4 MW motor. The fan is designed to generate a maximum speed of 225 m/s in the smallest section.

A special cooling station equipped with 4 compressors with a total cooling capacity of 6.4 MW achieves a minimum temperature of $-40\,°C$ in the smallest high-speed section and $-32\,°C$ in the other sections. The IWT has the unique feature for an installation of its size to simulate the pressure at an altitude of 7000 m, which can reproduce the actual flight conditions. This possibility is also very useful for studies of similarity laws for icing and for evaluating the influence of altitude on the shape of ice accretions. Figures 3.30a, b show typical tests performed in the IWT of CIRA.

The icing cloud generator (Spray Bar System or SBS), located 18 m upstream of the model, is designed to generate drops of water covering the envelope prescribed for icing certification and covering continuous and intermittent conditions. In addition,

(a) Ice accretion on the end of a wing in   (b) Icing on the air intake of a regional
a high-lift configuration   transport aircraft

Fig. 3.30 Typical tests in the IWT (© CIRA)

the SBS is able to generate large super cool drops and reproduce the freezing drizzle. For conventional aerodynamic tests, the spray module is replaced by a module containing three filters to achieve low turbulence intensities. A distinguished feature of icing wind tunnels is the need to ensure continued operation and maintenance of measuring instruments that determine cloud conditions. Reducing the uncertainty of these measurements and improving the process of cloud generation represent the two main areas of scientific and technological developments in icing wind tunnels.

3.2.4 Anechoic Chambers and Aeroacoustic Wind Tunnels

The **BETI (Bruit-Environment-Transport-Engineering) anechoic wind tunnel** at the PPRIME Institute in Poitiers, commissioned in 2013, is located in the premises of the University of Poitiers on the site of the Ecole Nationale Supérieure d'Ingénieurs de Poitiers (ENSI-Poitiers). It is a free-flow Eiffel type wind tunnel with a 90 m^3 plenum acoustically treated to reproduce free field conditions from 200 Hz (see Fig. 3.31). It allows the study and optimisation of the flow around obstacles and associated acoustic radiation. Its main characteristics are as follows: open-jet test section of 0.7 × 0.7 m^2 cross section and 1.5 m length, contraction ratio of 9, maximum speed of 60 m/s (216 km/h) and turbulence intensity less than 0.5%.

The BETI wind tunnel is equipped with various measuring systems: 3-axis traverse system and rotary table for the incidence of models, data acquisition system with up to 64 channels, 64 microphones, antenna system for the measurement of wall pressure fluctuations and velocity field measurement systems by hot wire anemometry, LDV or PIV. These devices allow the identification, analysis and control of noise sources (see Fig. 3.32). This facility is used for the aeroacoustic optimisation of elements or parts of ground vehicles, for fundamental studies related to aeronautics or for development of measurement techniques in aeroacoustics.

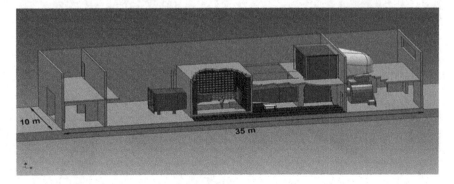

Fig. 3.31 Circuit of the BETI wind tunnel at the PPRIME Institute in Poitiers (© Institut PPRIME)

(a) Network of antennas for characterisation   (b) Measurement of the velocity field in the
    of aeroacoustic sources   vicinity of an upstream-facing step by PIV

Fig. 3.32 Plenum of the BETI wind tunnel (© PPRIME Institute)

The noise and wind installation of the **PROMETEE platform (Program and Test Facilities for Energy and Environment Transport,** see Fig. 3.33) represents an evolution of the existing aeroacoustic wind tunnel on the CEAT site (Centre of Aerodynamic and Thermal Studies).

This wind tunnel, used to study air jets in an anechoic environment, has two independent anechoic circuits powered by two compressors with the possibility of regulating the temperature and the stagnation pressure. Each compressor delivers a mass flow of 1 kg/s, with a total pressure of 3.5 bar and a temperature of up to 230 °C at the machine output. The air is then conditioned by air-to-air heat exchangers and motorised valves.

Fig. 3.33 PROMETEE Platform of the PPRIME Institute (CNRS—University of Poitiers—ISAE-ENSMA)

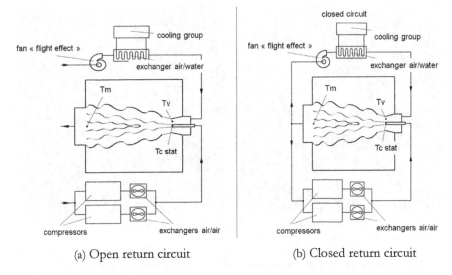

(a) Open return circuit   (b) Closed return circuit

Fig. 3.34 Operating modes of the "Noise and wind" facility of the PPRIME Institute (© PPRIME Institute)

The constitution of the plenum and large mass flow rate of the compressors makes it possible to create either two coaxial jets or a single jet. The jet diameter, a function of the Mach number, can reach 75 mm for Mach 1. A third circuit is used to simulate a "flight effect" by means of a 600 mm diameter jet at 55 m/s surrounding the central jet. This circuit is powered by a fan and includes a cooling system with a cooling capacity of 137 kW. As shown in Fig. 3.34, the facility can operate either in an open loop when the jets are discharged to the atmosphere, or closed loop through a system of vanes. The anechoic chamber, which has a volume of $12.6 \times 10.6 \times 7.85$ m^3, is lined with pyramidal foams ensuring the acoustic properties necessary for the studies considered. The wind tunnel is equipped with various measurement systems for probing the turbulent jet region (see Fig. 3.35a), the aerodynamic near field (see Fig. 3.35.b) and the acoustic field (see Fig. 3.36). The scientific objective is the development and validation of a jet noise theory based on hydrodynamic stability, allowing the development of simplified models for forecasting and control in 'free' and 'installed' configurations.

Two anechoic wind tunnels have been operated since the early 1980s at the **Acoustic Centre of the Laboratory of Fluid Mechanics and Acoustics (LMFA)** at École Centrale de Lyon. They result from the combination of open test section wind tunnels and a large acoustic chamber (see Fig. 3.37). The supersonic anechoic wind tunnel is fed continuously (air flow of 1 kg/s) by a 350 kW centrifugal compressor cooled by an air and water heater. At the outlet of the compressor, the air passes through a dryer, then by a soundproof and regulated heating box (72 kW), and by a set of particle filters. This facility is used for jet studies from nozzles whose diameter

(a) PIV, bi-planar, stereoscopic, high-speed   (b) Near-field' antenna with 48 microphones
    measurement system

Fig. 3.35 Instrumentation of the "Noise and wind" facility (© Institut PPRIME)

Fig. 3.36 Antenna with 18 microphones in the acoustic field of a turbulent jet in the «Noise and wind» facility of PPRIME Institute (© Institut PPRIME)

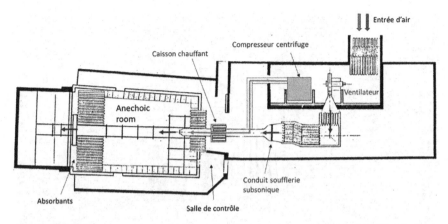

Fig. 3.37 Layout of the anechoic wind tunnel and the anechoic chamber of the LMFA of the Ecole Centrale de Lyon (© LMFA)

varies between 20 and 60 mm. The Mach number of the flow can reach 1.55 for a convergent nozzle with a diameter of 38 mm.

The subsonic anechoic wind tunnel is continuously fed (maximum mass flow 20 kg/s) by an 800 kW centrifugal fan. The air circuit is equipped with acoustic baffles, grids and honeycomb to homogenise the turbulence. In its terminal part, the circuit has a square section of 560 mm of side in which the level of turbulence does not exceed 0.5%. Different converging sections can be installed at the circuit outlet depending on the configuration studied. The maximum length of the test section is 8 m. The flow velocity corresponds to Mach 0.5 for an outlet section of 300 × 400 mm^2 and 0.8 for a circular section of 200 mm in diameter.

These two open-air wind tunnels are acoustically treated to cancel the noise of the air generation systems with respect to the measurements. They open-up into an anechoic plenum of 720 m^3 acoustically treated to reproduce free-field conditions from about 100 Hz (see Fig. 3.38). With motors at full power, the residual noise level in the anechoic chamber is less than 25 dB (A), much lower than the sound levels of the measured aeroacoustic sources.

The wind tunnels are equipped with instrumentations to characterise:

– near-field and far-field noise, using a set of antennas arranged in the anechoic chamber (117-microphone antenna array system, 13-microphone directivity antenna, coupled with a microphone acquisition system);
– the flow using steady and unsteady wall pressures, hot wire anemometry, LDV or PIV 3C-2D with high acquisition frequency.

The Acoustic Centre's wind tunnels are devoted to fundamental studies related to aeronautics, namely: subsonic and supersonic jet noise (see Fig. 3.39a), fan, high lift devices, landing gear, studies on parts or models of terrestrial vehicles

Fig. 3.38 The large anechoic chamber of the LMFA (© LMFA)

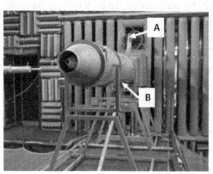

(a) Aeroacoustics of coaxial jets. (A) primary   (b) Aerodynamics and aeroacoustics of a land
flow, (B) secondary flow   vehicle model

Fig. 3.39 Examples of studies in the LMFA aeroacoustic wind tunnel (© LMFA)

(see Fig. 3.39b), or for metrological developments in aeroacoustics (optical methods, innovative sensors for wall pressure, methods and antennas for localisation of sources).

The jets can be coupled in order to exploit the acoustic characteristics of supersonic jets in flight conditions in the set-up shown in Fig. 3.39a. The flow from the supersonic nozzle acts as the primary jet and a secondary jet is generated by the subsonic nozzle in the conditions similar to that of the turbofan engines.

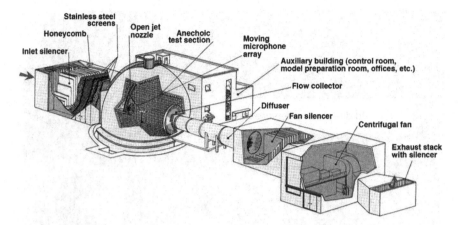

Fig. 3.40 General organisation of the ONERA CEPRA19 wind tunnel (© ONERA)

In Fig. 3.40, the **ONERA CEPRA19 anechoic wind tunnel** is equipped with an upstream settling chamber of 9×9 m^2 in cross section equipped with a dust filter, acoustic baffles, turbulence screens and a honeycomb. The nozzle has an exit diameter of 3 or 2 m depending on the test set up. The anechoic chamber (test section) has the shape of roughly a quarter of a sphere, with an internal radius of 9.6 m. The wind tunnel comprises of a collector-diffuser, a silencer for the fan and a centrifugal fan driven by a 7 MW motor. The maximum flow velocity is 130 m/s in a very good acoustic environment (minimisation of reflected noise) and very low background noise in the 200 Hz to 80 kHz frequency band. The large-diameter free-jet nozzle makes it possible to study 1/11 scale aircraft models as part of European noise reduction projects.

The wind tunnel is well suited for jet propulsion noise studies (maximum diameter equal to 0.3 m) whose temperature can reach 1150 K using a propane burner, with a secondary jet of temperature 500 K. It has conventional multi-holes probe and a three-component PIV measurement system. The measurements of the free-field acoustic directivity are carried out using two arcs of 12 microphones each, to characterise the acoustic levels in the "fly-over" directions (in the axis of the track) and "side-line" (on the lateral side of the runway) useful for the acoustic certification for take-off and landing. In addition, the wind tunnel provides the capabilities for identification of noise sources by acoustic antenna systems (see Fig. 3.41). These sensor arrays are instrumented with precision microphones, which can hold up to 260 microphones with a high-frequency sampling system up to 260 kHz.

The studies undertaken in CEPRA19 cover a wide variety of applications, where characterisation of in-flight jet noise and nozzle model tests are the most frequent (see Fig. 3.42). Other applications take advantage of the performance of this anechoic wind tunnel, in particular fans and the effects of motor installation, high lift devices and landing gear noise. CEPRA19 has also been used for the acoustics of land vehicles, high speed trains and cars.

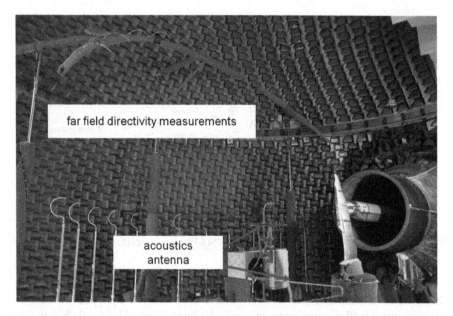

far field directivity measurements

acoustics
antenna

Fig. 3.41 Study of the motor installation effect on jet noise in the CEPRA19 wind tunnel (© ONERA)

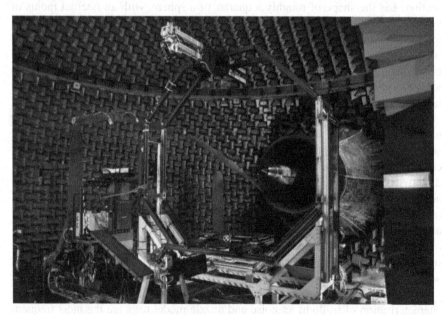

Fig. 3.42 3-components PIV measurement bench for the study of jet noise in the CEPRA19 wind tunnel (© ONERA)

3.2.5 Dual Purpose Aerodynamic and Acoustic Wind Tunnel

The **SAA (Soufflerie Aero-Acoustique) aeroacoustic wind tunnel at ISAE-SUPAERO** is a research wind tunnel with an anechoic chamber of a volume of $8 \times 9 \times 9$ m^3, for aeroacoustic studies in open jet mode, and the detachable test section of 1.8 m^2 and 6 m long can be used for conventional aerodynamic studies (see Fig. 3.43). The wind tunnel is driven by a fan powered by a 900 kW DC motor. In both the open jet and closed test section configuration the flow velocity can be continuously varied from 0 to 80 m/s (Mach number 0 to 0.24), corresponding to the speeds of civil transport aircraft in the landing and take-off phases.

The aeroacoustic circuit is equipped with acoustically treated walls and baffles limiting the propagation of the noise generated by the fan in the anechoic chamber. The plenum is equipped with several turbulence grids, a honeycomb and acoustically treated walls which, combined with the high contraction ratio of the convergent section, provide a turbulence intensity of less than 0.1% in an open or guided test section.

The side walls of the closed test section are either opaque or transparent removable panels adapted according to the tests. The anechoic chamber is equipped with noise

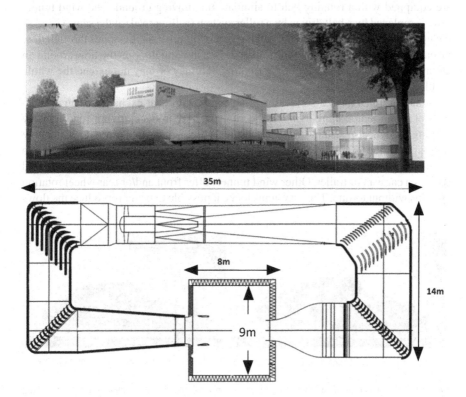

Fig. 3.43 The SAA aeroacoustic wind tunnel at ISAE-SUPAERO (© ISAE-SUPAERO)

attenuating cones with a cut-off frequency of 200 Hz and a background noise of less than 40 dB. A removable motorised floor allows quick and easy access to the models which is important considering the size of the anechoic chamber.

The SAA wind tunnel is devoted to fundamental or applied research related to aeronautics such as:

– aeroacoustics of aircraft and drones, for the reduction of noise from high-lift devices, landing gear, rotor noise, etc.;
– improved aerodynamic performance of aircraft and drones to test the efficiency of flow control devices, drag reduction, etc.

3.3 Wind Tunnels for Ground Vehicles

3.3.1 Specifications

Most of the wind tunnels dedicated to experiments on ground vehicles and railways are equipped with a rotating belt to simulate the moving ground. The wind tunnel floor is replaced by a belt driven by a roller system (rolling road) at the same speed as the upstream flow; i.e. equal to the vehicle speed. Figure 3.44 illustrates through flow visualisations in a water tunnel the influence of a moving floor on the flow around an automobile model. The width of the rotating belt may or may not extend the whole width of the vehicle.

Some facilities have a series of rolling roads with heat exchangers intended to control the air temperature in the test section and to reproduce driving conditions by varying the wheel-resistant torque. These tests are then carried out with the engine running and the gears engaged. In this case, the chassis dynamometer is used to simulate rolling of the tire on the road, to reproduce a particular additional load such as a caravan or trailer. Other wind tunnels offer front and/or rear wheel rotation systems without torque. These systems make it possible to better reproduce the flows

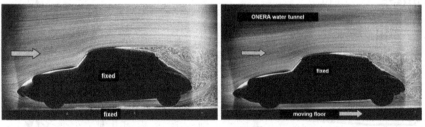

(a) Fixed floor   (b) Floor moving at the flow velocity

Fig. 3.44 Visualisation of the flow around a Citroën DS21 model in a water tunnel highlighting the ground effect (© ONERA)

around the wheels and in the wheel arches as well as the interactions with the under-
body flows and the vehicle wake. Autonomous temperature and flow measurement
systems can also be used to test and tune water, air or oil cooling systems. Water
supply and discharge systems designed to simulate the effect of rain with or without
wind and devices capable of reproducing snowy conditions are available in some
installations (see Sect. 3.2.2.).

3.3.2 Long Test Section Wind Tunnel

The Institut Aérotechnique (IAT) at Saint-Cyr-l'École is equipped with facilities for
studies in external or internal aerodynamics or related fields of aeroelasticity, with
applications in various sectors mainly in aeronautics, automotive, railway and civil
engineering applications.

The **Long Test Section Wind Tunnel or SVL (Soufflerie Longue Veine)** is a
closed circuit wind tunnel with a 15 m long test section equipped with a rolling
road of 6 m in length and working width of 0.4 m, recreating the ground effect with
rotating wheels (see Fig. 3.45). The tunnel was originally designed to study the drag
on different types of trains (see Fig. 3.46) and its dimensions also make it suitable
for civil engineering studies.

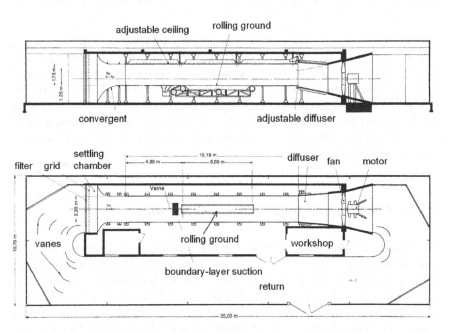

Fig. 3.45 Layout of the Long Test Section Wind Tunnel (SVL) (© IAT-CNAM)

Fig. 3.46 Study of a double deck TGV in the Long Test Section Wind Tunnel (SVL) (© IAT-CNAM)

While the 2.2 m width test section has solid walls, the inclination of the ceiling is adaptable in order to adjust for the effect of solid blockage over the entire length of the test section ensuring a zero-pressure gradient. The maximum speed of the rolling road corresponds to the maximum velocity of the flow in the test section; i.e. 40 m/s. A boundary layer suction device placed 0.5 m upstream of the moving floor reduces the thickness of the boundary-layer upstream of the model (17 mm without suction, 9 mm with suction).The tunnel is equipped with conventional steady and unsteady pressure measurement systems.

3.3.3 Wind Tunnels for Full Scale Automobile

The **S4 and S10 wind tunnels of IAT-CNAM** were designed to carry out full scale automobile aerodynamic tests while the S10 wind tunnel allowing also studies on 2/5 scale models. Special arrangements are provided to accommodate many types of vehicles or models adapted to the characteristics of wind tunnels and to meet the needs of studies or research in other fields such as aeronautics, railways, civil engineering or wind power. These closed-circuit wind tunnels have similar characteristics. Their test sections (5 m wide, 3 m high and 10 m long) have ventilated lateral and upper walls (see Fig. 3.47). These walls, with longitudinal slots, reduce the blockage created by

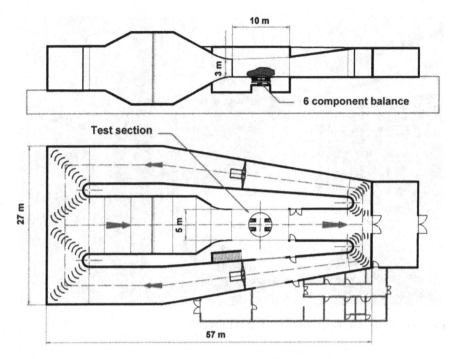

Fig. 3.47 Layout of the S10 wind tunnel circuit (© IAT-CNAM)

the presence of the model in the test section. The floors are equipped with a boundary-layer suction system to provide a better representation of the flow at ground level. The contraction sections are of ratios of 4 and 7.7 for the S4 and S10 wind tunnel with maximum velocities of 40 m/s and 55 m/s respectively.

These two wind tunnels are equipped with a turntable (4 m diameter for S4, 4.34 m for S10) incorporating a 6-component aerodynamic force balance, where the S4 wind tunnel has a rolling road. Figure 3.48 shows a competition car being tested in the S4 wind tunnel.

The S4 and S10 wind tunnels are also equipped with specialised equipment such as a water injection ramp for the simulation of rain and dirt.

The **S6 wind tunnel of the IAT-CNAM** is designed to perform aero-thermal studies on large vehicles (trucks, coaches and military vehicles). Its rectangular test section has an adjustable width of 4–6 m, by displacement of the side walls, a fixed height of 6 m and a length of 17 m (see Fig. 3.49). The maximum velocity of 20 m/s is generated by 27 fans located upstream of the test section.

Equipped with a rolling road with a maximum power of 315 kW and a maximum force at the wheel of 4000 N, the S6 wind tunnel is used for road simulation tests (endurance of alternators, cooling of braking systems). A 750 kW hot air generator allows for temperatures up to 55 °C to be reproduced in the test section. Sunshine is simulated by 180 infrared lamps generating 1.2 kW/m² (see Fig. 3.50). The wind

Fig. 3.48 Sport car in the S4 wind tunnel (© IAT-CNAM)

tunnel size makes it suitable for studies in other areas such as wind effect, sailing or low speed aeronautics.

In 2003, to meet the needs of the automotive industry, a highly sophisticated wind tunnel, with a power of 3.8 MW, was commissioned in Montigny-le-Bretonneux by an economic interest group (GIE S2A) bringing together PSA Peugeot-Citroën, Renault and the Conservatoire National des Arts et Métiers (CNAM). **The full-scale wind tunnel GIE S2A is a powerful mean to test full scale ground vehicles with all their real details.**

Of closed return type, the S2A wind tunnel achieves a maximum velocity of 240 km/h in a test section of 24 m^2 cross section, making it possible to study the aerodynamics and aeroacoustics of real vehicles (see Fig. 3.51). The wind tunnel is driven by a 3.8 MW motor and the open test section offers a good compromise between test section size and significant blockage effects at high skid angle. It also allows the positioning of acoustic measuring equipment on the side of the flow, out of the wind (see Fig. 3.52).

For the simulation of ground effects, the wind tunnel is equipped with a central rolling road between the wheels running at the flow speed and a belt under each wheel. It has a turntable to orient the vehicle at an angle with respect to the flow. The walls and ceiling of the plenum (semi-anechoic room) as well as part of the circuit are equipped with foam panels guaranteeing a background noise level below 58 dB, between a frequency band of 200–1600 Hz at 160 km/h. A three-axis traverse arm (left of Fig. 3.52) allows positioning the measurement probes throughout the whole volume surrounding the vehicle. A control circuit guarantees a temperature

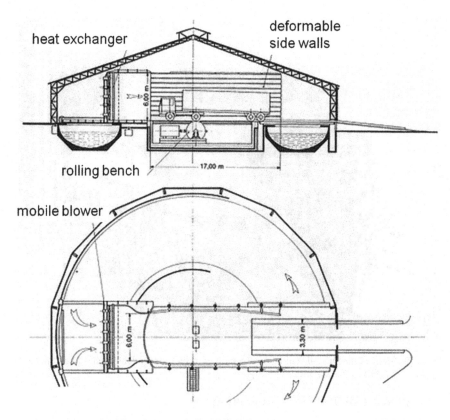

Fig. 3.49 Layout of the S6 wind tunnel (© IAT-CNAM)

between 15 and 25 °C necessary to perform tests throughout the year for aeroacoustic purpose in particular. The aerodynamic forces are measured by a six-component balance on which the vehicle rests by means of its wheels and mounting cylinders (see Sect. 8.2.4).

With its rolling road and boundary-layer suction system to approach the actual conditions of the vehicle on the road, it is one of the most optimised wind tunnels of this type in Europe, if not in the world. It is used primarily to study vehicles aerodynamics and in particular for drag reduction, in direct link with the objectives of reducing fuel consumption and CO_2 emissions. The second use of the wind tunnel is the control of the aerodynamic noise sources, with a view to improving passenger comfort.

A PIV system, pressure transducers, anemometers and visualisation tools allow a fine flow analysis. Figure 3.53a shows visualisation test with smoke injected upstream of the model.

For acoustic measurements, holography and beam-forming techniques are used, including local measurements on mirrors and even mannequins to measure the noise perceived by the driver and passengers (see Fig. 3.53b). Beam-forming is an acoustic

Fig. 3.50 Study of the internal ventilation of a military vehicle under controlled sunlight in the S6 wind tunnel (© IAT-CNAM)

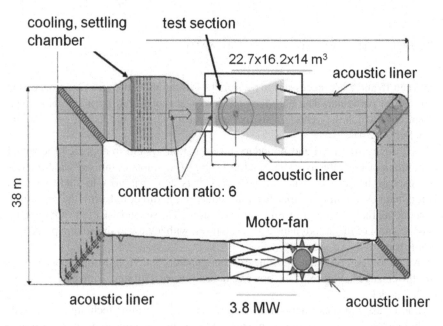

Fig. 3.51 Layout of the S2A wind tunnel at Montigny-le-Bretonneux (@ S2A Wind Tunnel)

Fig. 3.52 Test chamber of the S2A automobile wind tunnel (@ S2A Wind Tunnel)

(a) Smoke visualisation

(b) Acoustic measurements with mannequin equipped with ear-level microphones

Fig. 3.53 Test on a Renault CLIO IV in the S2A wind tunnel (© S2A Wind Tunnel)

mapping technique that requires a large number of microphones (at least 60) arranged on a plane antenna (grid) placed parallel to the object to be studied. The dimensions of the test room and the background noise level also allow aeroacoustic tests for building and aeronautics.

On the same site, a smaller wind tunnel is used for tests on 2/5 scale models. The maximum velocity of the flow is 240 km/h. A six-component balance allows measurement of the aerodynamic forces. In addition to the automobile applications, it is also used for tests on bicycle and drones.

Founded in 2002, **Aero Concept Engineering (ACE) in Magny-Cours** conducts numerous aerodynamic studies, particularly in the field of motorsport. It is equipped

with a closed circuit wind tunnel with a test section which is 2.3 m wide, 2.2 m high
and 4.75 m long (see Fig. 3.54). The contraction ratio is 7 and a turbulence intensity
of 0.1% is attained at a maximum velocity of 40 m/s with a 250 kW fan. The wind
tunnel can be equipped with a rolling road 3 m long and 1.5 m wide that can travel at
a maximum speed of 40 m/s for model scales ranging from 25 to 50%. The upstream
boundary layer can be aspirated. The wind tunnel has a 6-component main balance
and other 3-component balances (fin, wheels). In addition to the force measurements,
the wind tunnel instrumentation allows surface oil flow visualisations, wall pressure
and PIV measurements. Figure 3.55a shows a prototype model tested with a rolling
road wider than the model which is then held by a mast mounted on the roof with
lateral supports for measuring the forces on the wheels. The rolling road can be
replaced by a 1.5 m diameter turntable in order to measure skid forces on many other
configurations such as bicycle, motorcycle, aircraft, building, etc. Figure 3.55b shows
a test on a race motorcycle for the identification of the most optimised aerodynamic
position.

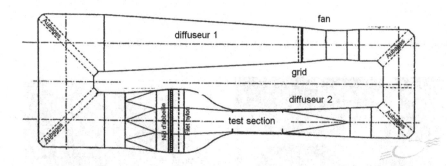

Fig. 3.54 Layout of the ACE wind tunnel at Magny-Cours (© Aero Concept Engineering)

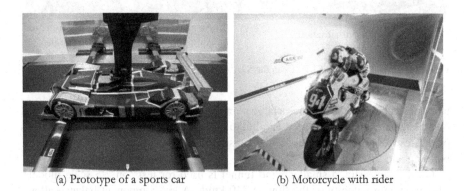

(a) Prototype of a sports car   (b) Motorcycle with rider

Fig. 3.55 The ACE wind tunnel at Magny-Cours (© Aero Concept Engineering)

3.4 Water Tunnels

3.4.1 General Description

Although strictly speaking not an aerodynamic means, water tunnels have been and are still widely used to study low velocity flows around aerodynamic objects. The justification for their use is due to the fact that at low speeds air can be considered as an incompressible fluid which is the case of water. Very low velocity water flows (close to 1 m/s) lend themselves very well to visualisations by injections of dye or fluid tracers, which make it possible to highlight separation and vortex formation. These tracers can be injected either upstream of the model or through orifices at the model wall. Other more sophisticated techniques can be used (see Sect. 7.3). The main disadvantage of the water tunnel is the low Reynolds number achievable, this inconveniency being secondary in the study of largely separated regions whose formation and development do not depend much on the Reynolds number. Thus, water tunnels have been extensively used for the study of vortex formations on delta wings, base flows and three-dimensional separated structures in general. Three types of arrangement exist:

The closed return circuit water tunnel has an architecture similar to that of a closed return circuit wind tunnel, the circulation of water being ensured by a pump, such as the ONERA Thales water tunnel (see Fig. 3.56). This type of facility is well

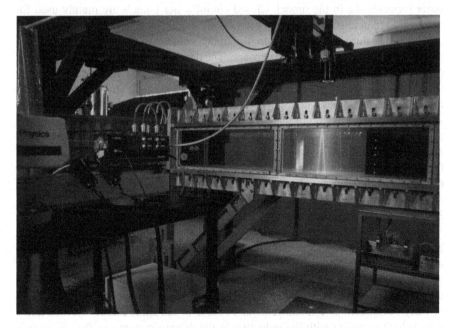

Fig. 3.56 Test section of the ONERA Thales water tunnel (© ONERA)

Fig. 3.57 The water flume of the ONERA Lille centre (© ONERA)

suited to local measurements by PIV, for example. On the other hand, visualisations by injection of coloured fluid are difficult because of the rapid contamination of the water recirculating in the tunnel. Closed circuit water tunnels are mainly used for real hydrodynamic studies, such as cavitation.

The water flume, or hydrodynamic channel, consists of a long channel in which the model is towed by an external carriage. The device provides a controlled velocity over a time depending on the length of the tunnel. The water flume at ONERA in Lille is 22 m long, 1.5 m wide and 1.5 m deep (see Fig. 3.57). This facility was used for aerodynamic studies with models suspended on its trolley. Hydrodynamic impact studies were also carried out thanks to a tilting hydraulic actuator, mounted on the trolley and at the end of which instrumented models were fixed. In the early 1980s, a motor pump was installed to generate a circulation of water. In the 2000s, the water flume was used for the study of flapping wings (applications to micro-drones), either on trolley, or in water circulation mode. Force measurements and visualisations by dye injections were done for unprecedented Reynolds numbers ranging from 300 to 30,000. Sail wake tests are also performed with the use of PIV.

The vertical water tunnel works under the effect of gravity. After filling a tank and water tranquilisation, opening a downstream valve drains the tank and establishes a flow in the test section, (see Fig. 3.58). This device, which has the advantage of its simplicity of design and use, lends itself well for flow visualisations as the water contaminated by dyes is immediately evacuated. On the other hand, it is more difficult to carry out quantitative measurements because of the short useful test time and a flow velocity varying with the reduction in the driving pressure as the reservoir is emptied.

Fig. 3.58 The gravity driven
water tunnel TH2 of
ONERA (© ONERA)

3.4.2   Low Speed Water Tunnel

In 1996, ONERA Lille set up a vertical **low-velocity Hydrodynamic Tunnel
(THBV)** to study three-dimensional aerodynamic phenomena and unsteady flows
(see Fig. 3.59). This facility has a square test section of 300 mm^2 and length of
1.5 m, the maximum continuous speed being 1 m/s. The test section is equipped
with glass windows of Schlieren and interferometric quality with an area of 265
× 465 mm^2. They are interchangeable in both perpendicular directions and in two
sections of the test section. Model supports, at the wall or between walls, are possible. This facility is equipped with a plate separating the contraction section into
two channels. Using a system of double pumps, it is possible to produce flows with
different velocity at the test section inlet. Sting balances and a specific balance for
sheared flows complete the instrumentation set which includes dye visualisations by
laser sheet or Schlieren with thermal marking (see Sect. 7.5.1), local measurements
by hot film anemometry, PIV or LDV.

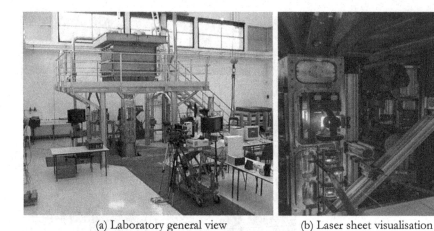

(a) Laboratory general view   (b) Laser sheet visualisation

Fig. 3.59 Low speed water tunnel at ONERA, Lille (© ONERA)

3.4.3   Hydrodynamic Channel for Polarographic Measurements

The polarographic measurement method is based on the diffusion properties of certain oxidising-reducing pairs in aqueous solution. It requires the use of electrochemical cells (polarographic) formed of a measuring electrode (cathode) and a counter-electrode (anode) separated by an electrolyte. Closing the circuit between the anode and the cathode creates an oxy-reduction reaction that generates an evolution of the chemical components at the measuring electrode. This evolution provides access to instantaneous surface velocity gradients and hence to local skin friction. The polarographic solution must have a Newtonian behaviour to reproduce the linearity of the viscous sub-layer in the vicinity of the wall. The chemical reaction must be almost instantaneous and reversible, to limit the consequences of chemical kinematics on the measurement.

The **hydrodynamic channel of the Université polytechnique des Hauts-de-France** (previously Université de Valenciennes et Hainaut-Cambrésis), used for polarographic measurements, consists of a plenum, a convergent, a horizontal test section, a heat exchanger and a fan/propeller connected to an engine (see Fig. 3.60). The elements of the channel are mainly made of polymethyl methacrylate (PMMA) and polypropylene, chemically inert and electrically neutral materials. The use of a DC motor ensures the non-perturbation of electrochemical measurements.

The plenum chamber, (1), with an inner diameter of 1.4 m comprises a honeycomb as a flow straightener, a second filter being placed downstream of the test section. The contraction ratio of the convergent, (2), is 17 and the turbulence intensity at the centre of the test section is less than 1.5%. The test section, (3), with a square cross-section of 0.09 m^2 and a length of 1.2 m, has transparent removable walls on all four sides. Part (4), of length 0.65 m prolongs the test section. The fluid is set in

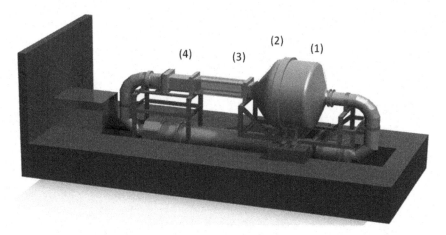

Fig. 3.60 The hydrodynamic tunnel of the Université polytechnique des Hauts-de-France

motion by a DC motor with a power of 52 kW whose nominal speed is 940 rpm. These motor drives a propeller placed in the return circuit to achieve velocities in the test section close to 4 m/s.

Chapter 4
Transonic Wind Tunnels

4.1 Definition of the Transonic Regime

As the air flow around an aircraft is not uniform, locally supersonic zones may exist even under subsonic upstream flow condition, this defines the transonic regime, a condition where both subsonic and supersonic regions coexist. This situation significantly modifies the local behaviour of the airflow. Transonic phenomena were responsible for the numerous failures and crashes during the first attempts to cross the sound barrier, that is, to fly at a speed greater than that of the sound (Mach number greater than 1). Similarly, in a supersonic upstream flow, there could be subsonic regions that influence the overall flow if they are significant in volume. The aerodynamic phenomena induced by the coexistence of subsonic and supersonic regions are complex and justify the definition of a so-called transonic flight regime. It is considered that transonic phenomena can be significant in flight conditions between Mach 0.7 and Mach 1.2. This is currently the cruise condition for all commercial jets in operation.

Transonic wind tunnels, in which freestream velocities close to the sound velocity are reached, have been the subject of many developments because of the criticality of the transonic phenomena, their strategic importance, and the particular design problems they address. As shown in Fig. 5.1 of Chap. 5, in the vicinity of Mach one, a very small variation of the test section area accelerates the flow from subsonic to supersonic with drastic changes in its behaviour. Accordingly, the presence of the model in the test section causes a local decrease in the cross sectional area available to the fluid which then becomes a throat, so this results in a sonic blockage of the flow due to the Mach number being sonic at the model location. The un-choking of the flow can be achieved by various techniques that were intentionally kept secret in the 1950s.

© Springer Nature Switzerland AG 2020
B. Chanetz et al., *Experimental Aerodynamics*,
Springer Tracts in Mechanical Engineering,
https://doi.org/10.1007/978-3-030-35562-3_4

4.2   Blockage Reduction and Flow Un-Chocking

4.2.1   Perforated or Slotted Walls

A first and quick solution is to equip the test section with perforated walls, with the holes in communication with each other through a plenum chamber, a portion of the flow being ducted through this chamber as shown schematically in Fig. 4.1.

This results in a virtual widening effect of the section, avoiding the creation of a sonic throat. Figure 4.2 shows a setup for the study of the base region of a propulsive

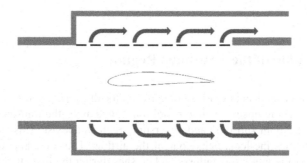

Fig. 4.1 Un-chocking of the flow in a transonic section

Fig. 4.2 An afterbody configuration in the S3Ch wind tunnel equipped with perforated walls (© ONERA)

Fig. 4.3 ETW wind tunnel
equipped with slotted walls.
Downstream view showing
the plenum (© ETW)

afterbody installed in the S3Ch wind tunnel at the ONERA Meudon centre equipped
with perforated walls.

In a variant of this method, the perforations are replaced by longitudinal slots
intended to disturb the flow slightly less than holes (see Fig. 4.3). However, either
holes or slots do not guarantee a reproduction of the flow as expected around a vehicle
in a free atmosphere. In addition, the holes behave like resonant cavities producing
acoustic radiation capable of triggering a premature laminar to turbulent transition of
the boundary layer on the model. Also, slotted or perforated walls can pose serious
difficulties for numerical simulations that tend to take into account all the space
around the model, including the walls of the wind tunnel.

4.2.2   Adaptive Walls

A more recent technique consists of using adaptive walls for the test section which
deform into the shape of a stream surface over the model when place in the confined
straight test section. The deformation of the wall first ensures that the flow is not
choked by a local increase of the cross section and secondly compensates for the
disturbances created by the model so as the model is tested in conditions similar to
that in the absence of solid walls. The correction method consists of shaping the walls

such that their curvatures are streamlines of the flow tending towards a uniform state at infinity. The principle of adaptive walls is most often applied to two-dimensional flows where only the upper and lower walls of the test section being adapted.

In practice, the upper and lower walls of the transonic test section consist of a flexible steel sheet that can be deformed under the action of jacks, as shown in Fig. 4.4, and equipped with pressure tapings. Adaptation is based on the comparison of the measured pressure distribution with that of a uniform flow which would exist in the far field, in the presence of the model. This fictitious flow in the far field is determined with the wall of the test section as a boundary condition and in theory it is a streamline of the flow. This calculation can be done by an approximate analytical method (one is in principle far from the model), or by solving the Euler or Navier-Stokes equations. The resulting pressure distribution being generally distinct from that of the measured wall pressure distribution, the adaptation procedure consists of deforming the wall until the measured and calculated pressure distributions coincide. The calculation of the external flow can be coupled to an optimisation algorithm using the least squares method and a Gauss-Newton algorithm to find the optimal shape. The process converges very rapidly, two iterations being most often sufficient. Then, the conditions in the test section can be considered as those existing in an infinite domain where the flow becomes uniform. The implementation of the adaptive walls requires a complex system coupling a set of jacks to deform the walls and a computer calculating the required shape according to the desired Mach number, while correcting for the blockage effect induced by the model.

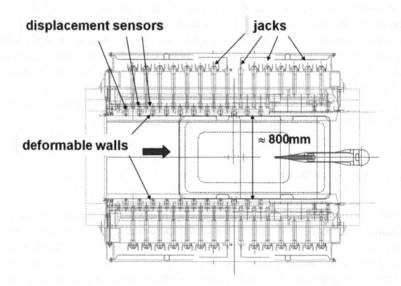

Fig. 4.4 Adaptive walls of the S3Ch wind tunnel at ONERA, Meudon (© ONERA)

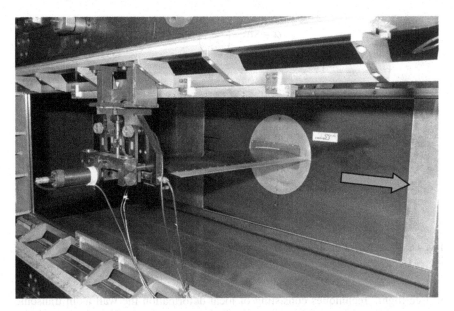

Fig. 4.5 Set-up for measurement of forces and moments on a profile by a 6 component balance in the test section of the ONERA S3Ch wind tunnel equipped with adaptive walls (© ONERA)

Figure 4.5 shows a setup used for the measurement of forces and moments on a wing profile in the test section of the S3Ch wind tunnel equipped with adaptive walls. The setup is equipped with a 6 components balance mounted on a side wall, not shown in the photograph.

4.2.3 Reflection of Disturbances

Another difficulty encountered in the design of a transonic wind tunnel operating in a slightly supersonic regime (upstream Mach number between 1 and 1.3) results from the propagation of supersonic disturbances. Indeed, the disturbances produced by the model propagate along Mach lines whose slope relative to the upstream velocity vector is equal to the Mach angle:

$$\alpha = \sin^{-1}\left(\frac{1}{M}\right) \approx 90°$$

This slope being close to 90° near Mach 1, these waves also reflected on the walls of the test section at an angle also close to 90° can impact on the model (see Fig. 4.6). This is a very serious issue to which there is little remedy, except testing very small models compared to the dimensions of the test section or to test these models

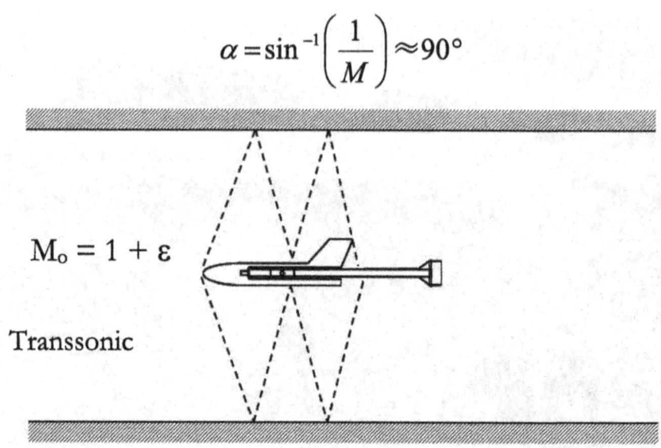

$$\alpha = \sin^{-1}\left(\frac{1}{M}\right) \approx 90°$$

$M_o = 1 + \varepsilon$

Transsonic

Fig. 4.6 Reflection of disturbances on the top and bottom walls of the test section in supersonic flow

in free flight. Techniques consisting of local deformation the wall as in transonic tunnels to compensate for the wave reflections are difficult to implement because of requirements for local curvature in the wall to achieve supersonic speed.

4.2.4   Double Throat Diffuser

Transonic wind tunnels are often equipped with a throat located downstream of the test section or sometimes in the diverging section, a second throat. It normally constitutes of movable walls made out of adjustable plates, creating a convergent-divergent supersonic nozzle and starting-up the tunnel involves the passage through two flow regime of interest. First by setting the Mach number in a starting supersonic nozzle depending on the ratio of the local section and the throat (see Sect. 5.1) and an adjustment of the section of the second throat makes it possible to fix the Mach number in the subsonic part of the test chamber with a very high precision (a thousandth). The supersonic flow in the diverging part of the downstream throat prevents the upstream propagation of disturbances generated downstream of the test section. The flow in the test section is thus isolated from slight fluctuations in the operation of the wind-tunnel fan. The higher power required to achieve a slightly supersonic flow in the converging-diverging nozzle of the second throat must be accounted for. Figure 4.7 shows the test section of the S8Ch transonic wind tunnel equipped with a second throat, at ONERA, Meudon.

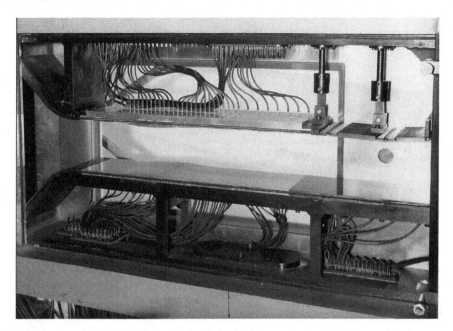

Fig. 4.7 Test section of the ONERA S8Ch wind tunnel with a second throat (© ONERA)

4.3 Typical Transonic Wind Tunnels

4.3.1 Very Large Transonic Wind Tunnel

The very large **S1MA transonic wind tunnel at ONERA, Modane-Avrieux** shown by the 3D assembly drawing in Fig. 4.8, is equipped with two 15 m diameter counter-rotating fans with 10 and 12 blades with adjustable pitch. These fans are driven by Pelton turbines with a capacity of 44 MW each.

The turbines that drive the fan are supplied with 10 million cubic meters of water per year from several reservoirs managed by EDF (Electricity of France). From the dam above Avrieux in the Alps, a waterfall of 840 m high provides the energy to drive the wind tunnel through a penstock sluice where the flow rate can reach 15 m³/s. The cooling of the wind tunnel is ensured by exchange of air with the atmosphere through air vents. The test section has a diameter of 8 m and a length of 14 m, making S1MA the largest transonic wind tunnel in the world. The wind tunnel has 3 interchangeable test sections mounted on trolleys to ensure ease of model change-over and rapid turnaround for testing various configurations (see Fig. 4.9).

One of the test sections may be equipped with solid or slotted walls/liners. The Mach number is continuously adjustable from 0.05 to 1 by varying the rotation speed of the fans from 25 to 212 rpm. The total pressure is approximately equal to the local atmospheric pressure; i.e., 0.9 bar, the stagnation temperature being between 258

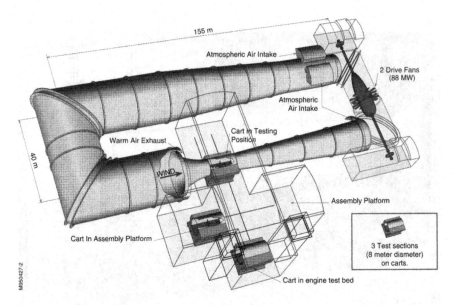

Fig. 4.8 The S1MA transonic wind tunnel at ONERA, Modane-Avrieux (© ONERA)

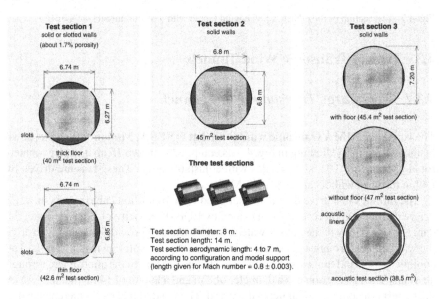

Fig. 4.9 Different configurations of the test section of the ONERA S1MA transonic wind tunnel (© ONERA)

and 333 K. Depending on the Mach number and the external temperature a Reynolds number of about 10^7/m can be attained at Mach 1.

The wind tunnel has a very complete set of means to quantity the flows including measurement of forces, steady and unsteady pressures (by PSP in particular), and for qualitative measurements, infrared thermography for heat fluxes, visualisations by smoke or viscous coating and the detection of the transition by naphthalene sublimation. A diverse range of rigs are available for testing propulsive jets, helicopter rotors, propellers or air intakes.

The wind tunnel also allows for icing tests and probing the flow around the model. Among the tests carried out in S1MA, are tests on a complete airplane model with aerodynamic forces measurement by 6-component balance and also measurements on helicopter rotors, on full-scale missiles, on interferences nacelle/pylon/wing (with nacelle equipped with TPS, see Sect. 2.7), on load jettisoning, on the control of the laminar flow with boundary layer suction, air intake, etc. Figure 4.10 shows a set-up of a missile fired from a combat aircraft. In this setup, the model of the missile is held by a robot arm which is operated remotely by a computer which positions the missile and measures the forces to which it is subjected.

Fig. 4.10 Test of missile separation under a combat aircraft in the ONERA S1MA wind tunnel (© ONERA)

4.3.2 Transonic Wind Tunnel for Research

The **S3Ch wind tunnel at the ONERA Meudon centre** is a research wind tunnel whose dimensions are appropriate to the studies of part configurations such as wing profiles, after-bodies, wing-nacelle sets, etc. Its Mach number range extends from 0.3 to 1.2. Its total pressure is close to the atmospheric pressure and total temperature being in the range of 290–330 K. The average turbulence intensity in the upstream part of the working section is 0.15%.

The tunnel is operated by a two-stage power unit with a power of 3.5 MW and a compression ratio of 1.25. The test section shown in Fig. 4.11 has a rectangular cross section of 0.76×0.80 m^2 and a length of 2.2 m. It can be equipped with rigid top and bottom walls, perforated, or with deformable and adaptive walls (see Sect. 4.2.2), which makes it a unique facility. The side walls include optical grade windows for optical flow diagnostics. The wind tunnel is equipped with an axial model support with internal balances allowing a variation of incidence angles or sideslip angles of 60°, without significant displacement of the model from the centre of the test section. It is possible to mount the models on the wall of the test section with external or internal balance and variable incidence over 360°. A compressed air supply at a rate of 2 kg/s, under a pressure of 40 bar and a temperature of 300 K can simulate propulsion jets or supply a TPS (see Fig. 4.12). The wind tunnel also has a 90 kW heater, to provide hot air flow at a rate of 0.5 kg/s, to simulate a propulsion jet at a temperature of 600 K.

The wind tunnel allows implementation of flow diagnosis techniques for steady and unsteady pressure measurements using PSP and PIV, two- and three-components LDV for measurement of the velocity field, as well as various visualisation methods. It is also widely used for laminar to turbulent transition control studies.

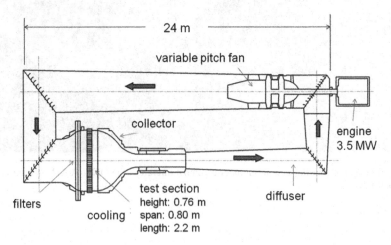

Fig. 4.11 Circuit of the S3Ch transonic wind tunnel at ONERA, Meudon (© ONERA)

Fig. 4.12 Test section of the ONERA S3Ch transonic wind tunnel with a model of a wing portion of an A340 powered by TPS. The model is mounted on the sidewall with adaptive bottom walls (© ONERA)

4.3.3 Transonic-Supersonic Wind Tunnel

The **S2MA at ONERA, Modane-Avrieux** is a wind tunnel with the capability of operating up to Mach 3 with a range covering 0.3–3.1. The tunnel is of a continuous and return type and it is operated by a compressor driven by water turbines that generates a power of 55 MW, similar to the driving system adopted for the S1MA wind tunnel (see Fig. 4.13).

In subsonic-transonic configuration, the test section has a rectangular cross section 1.77 m high and 1.75 m wide, the upper and lower walls being perforated. In supersonic configuration, the test section has a rectangular cross section of 1.95 m high and 1.75 m wide. As shown in Fig. 4.14, the supersonic nozzle has an asymmetric contour; the movable bottom wall can be translated to change the height of the throat and the geometry of the diverging section so as to achieve a continuous Mach number variation from 1.5 to 3.1. The contour is calculated to ensure a Mach number nearly uniform in the test section for each position of the movable wall.

The total pressure can vary between 0.25 and 2.5 bar during transonic operation, and between 0.25 and 1.75 bar in the supersonic configuration with the stagnation temperature below 313 K. Figure 4.15 shows the operation range of the wind tunnel in terms of both Mach number and Reynolds number. It is important to note that here the experiments can be conducted at conditions close to cruise conditions of regional jets.

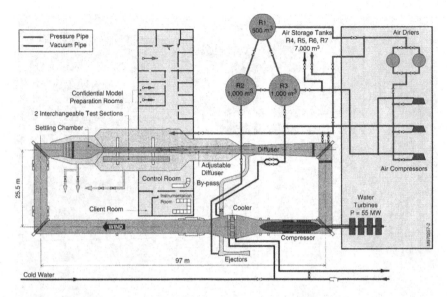

Fig. 4.13 Organisation of the ONERA S2MA trisonic wind tunnel (© ONERA)

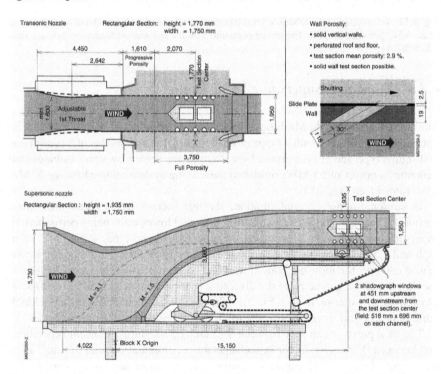

Fig. 4.14 Supersonic test section of the ONERA S2MA trisonic wind tunnel (© ONERA)

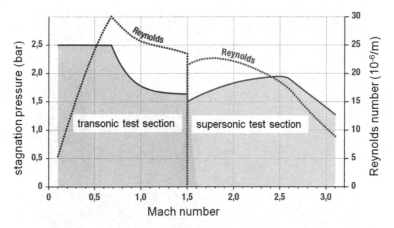

Fig. 4.15 Operational range of the ONERA S2MA trisonic wind tunnel (© ONERA)

The wind tunnel is equipped with a complete means for measuring and qualifying the flow: steady and unsteady pressure measurements, aerodynamic forces and moments by internal balance or mounted on a wall, flow visualisations, PIV and PSP. It is also equipped with techniques for the simulation of hot or very hot (pyrotechnic) propellant jets. Tests performed in S2MA include: air intake, load jettisoning, dynamic stability, flutter, nacelle/pylon/wing interference and transverse jet studies. Figure 4.16 shows the set-up for measurements on a business jet model: note the upper and lower perforated walls.

4.3.4 A Cryogenic and Pressurised Wind Tunnel: The European Transonic Wind Tunnel

The European Transonic Wind tunnel (ETW) in Cologne is a cryogenic and pressurised wind tunnel (see Sect. 2.2) composed of a closed-loop aerodynamic circuit contained inside a thermally insulated and pressurised stainless steel housing (see Fig. 4.17). The wind tunnel operates on nitrogen whose thermodynamic properties are near identical to those of air (same value of the ratio of specific heats γ). A compressor with a maximum power of 50 MW ensures the circulation of the nitrogen gas. To reach the desired low temperature and to compensate for the heat generation due to the viscous friction in the flow, liquid nitrogen at a temperature of 110 K is continuously injected into the return circuit, after the first turning vane, downstream of the diverging section and upstream of the fan, through four injectors with about 230 spray nozzles. This liquid nitrogen vaporises instantly to form the cold flow. The corresponding nitrogen gas outlet is located in the opposite return circuit upstream of the plenum and is controlled by valves. This exhaust is adjusted accordingly with the rotation speed of the fan and the mass flow rate of liquid nitrogen to keep the pressure

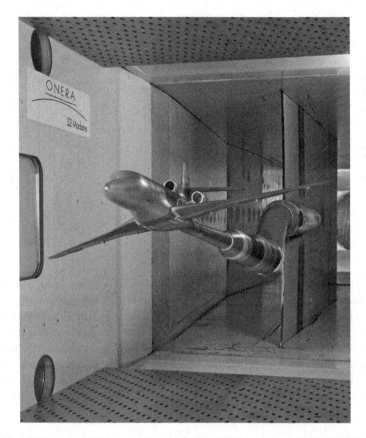

Fig. 4.16 A business jet model mounted in the ONERA S2MA wind tunnel (© ONERA)

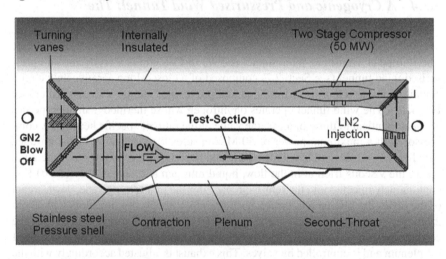

Fig. 4.17 Circuit of the ETW wind tunnel in Cologne (© ETW)

constant inside the wind tunnel. Downstream of the test section there is an adjustable second throat, designed to minimise flow disturbances propagating upstream and to provide highly accurate Mach number control during tunnel operation at Mach numbers between 0.7 and 1.0.

The test section has a height of 2 m, a width of 2.4 m and length of 9 m. The wind tunnel has a slotted transonic liners and a supersonic nozzle designed for the tests at Mach number of 1.35. Based on the dimensions of the test section a typical full-span aircraft model would have a wingspan of about 1.6 m, while half models, which are mounted with their half fuselage from the ceiling of the test section, can have a semi-span of about 1.3 m. Figure 4.18 shows a model of an Airbus A320 in the ETW wind tunnel test section. Optical access to the test section for various cameras and light sources is provided through 90 special windows in all four walls.

The specifications of the wind tunnel are: Mach number from 0.15 to 1.35, total temperature of 313–110 K, total pressure of 1.15–4.5 bar. The combination of the

Fig. 4.18 Model of an Airbus A320 in the ETW wind tunnel in Cologne (© ETW)

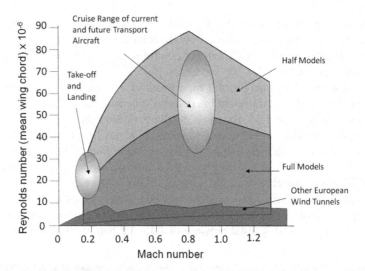

Fig. 4.19 Operating domains of the ETW wind tunnel in Cologne (© ETW)

size of the working section and the environmental conditions helps in attaining similar conditions achieved during cruise conditions which makes this tunnel the most strategic for the aerospace industry.

The operating ranges with respect to each Mach number can be summarised into an overall ETW performance envelope expressed as Reynolds number against Mach number in Fig. 4.19. Pure Reynolds number or aerodynamic loads variations at constant Mach number can be achieved by adjusting the total pressure and temperature accordingly. Testing at ambient temperature allows data comparison with conventional low- or high-speed tunnels.

Up to the Mach number of 0.7 (Mach number 1 if the second throat is fully open), the flow speed is controlled by the fan speed. From the Mach number of 0.7–1, the second throat is used to control the Mach number of the flow in the test section. Beyond the Mach number of 1, the supersonic nozzle is the main element controlling the flow speed. With full-span aircraft models a Reynolds number of about 50 million can be reached, and when a larger ceiling mounted half (semi-span) model is used then Reynolds numbers of up to 85 million can be achieved. These Reynolds numbers are based on the mean aerodynamic cord of the respective wing. The upper Reynolds number limit is obtained at the minimum temperature and at the first reach of either the maximum pressure or the maximum drive power.

To ensure the operation in the cold environment, the ETW team has developed a transportable model interchange system based on a cart containing the model and its support, the top wall of the test chamber, the access hatch and the instrumentation box, the full assembly being moved by a carrier. The models are prepared in two halls: one with ambient air and a second "dry" hall filled with dry air at room temperature whose dew point is kept low enough to prevent icing and ice formation on the cold model. The two halls are separated by a transfer air locked, where the model and its trolley are purged of all moist air before entering the dry hall. Two of such model cart assemblies are available to allow rigging work on models and testing to be performed in parallel.

Chapter 5
Supersonic Wind Tunnels

5.1 Convergent-Divergent Nozzle

To begin with, the generation of a supersonic flow requires a convergent duct and an upstream/downstream pressure ratio sufficient to achieve sonic velocity at the duct minimum section or throat. Downstream, the velocity of the flow continues to increase and becomes supersonic if the section of the duct increases (Hugoniot's relationship) and if a low enough pressure is maintained. Such a convergent-divergent nozzle is sometimes called a de Laval nozzle named after its inventor, the Swedish engineer Gustaf de Laval. The laws of thermodynamics and fluid dynamics make it possible to establish the following relation, usually referred as the isentropic relation for flow in converging-diverging (CD) nozzles, which provides a critical area relation between the cross section A of the duct and the Mach number, M:

$$\frac{A}{A_c} \equiv \Sigma(M, \gamma) = \left(\frac{2}{\gamma + 1}\right)^{\frac{\gamma+1}{2(\gamma-1)}} \frac{1}{M}\left(1 + \frac{\gamma - 1}{2}M^2\right)^{\frac{\gamma+1}{2(\gamma-1)}}$$

where A_c is the minimum cross section of the nozzle or throat. The curve in Fig. 5.1 shows a rapid decrease in the ratio of the cross section area as a function of the Mach number in subsonic regime, then a flat evolution on both sides of the Mach number close to 1 in transonic regime, before a further rapid increase. The very progressive evolution of the test section area in the transonic regime is one of the sources of the difficulties encountered in the design of transonic wind tunnels (see Sect. 4.2).

Figure 5.2 presents the evolution of the pressure and the temperature (relative to the stagnation conditions) as a function of the Mach number; it shows that the acceleration of a supersonic flow causes a rapid decrease in pressure and temperature. Therefore, the establishment of the flow requires the use of a powerful fan unit or compressor in order to maintain such a large pressure difference between the upstream and downstream parts of the test section. A solution adopted in high-Mach number wind tunnels (beyond Mach 3) is to expand compressed air stored

© Springer Nature Switzerland AG 2020
B. Chanetz et al., *Experimental Aerodynamics*,
Springer Tracts in Mechanical Engineering,
https://doi.org/10.1007/978-3-030-35562-3_5

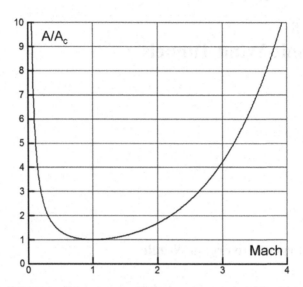

Fig. 5.1 Evolution of the Mach number according to the cross-section area ratio

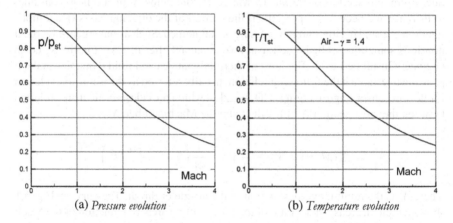

(a) *Pressure evolution*   (b) *Temperature evolution*

Fig. 5.2 Isentropic expansion in a nozzle

in a tank. The pressure ratio between upstream and downstream can be increased by connecting the wind tunnel diffuser to a large volume tank where a very low pressure is established before testing. Such an installation is called an intermittent wind tunnel, the time during which the supersonic flow is established being limited by the capacity of the compressed air storage or by the volume of the vacuum tank and also by the quality of the diffuser. Indeed, a good diffuser can recover a higher pressure after the minimum reached in the test chamber, hence the ability to maintain the flow with a higher pressure in the vacuum tank.

Due to the cooling of the air during the isentropic expansion in the nozzle, the water vapour present in the ambient air condenses in the form of condensation shocks resulting from the abrupt passage to the liquid state. This results in a perturbation of the flow from the transonic regime which requires equipping the supersonic wind tunnels with a dryer in order to eliminate humidity from the ambient air. In transonic/supersonic wind tunnels, the compressed air is desiccated (dried) before storage in the tanks. In high supersonic wind tunnels, it is also necessary to heat the air due to the cooling produced by the adiabatic expansion in the nozzle in order to avoid liquefaction.

5.2 Defining of the Contour of a Supersonic Nozzle

To obtain a uniform flow in the test section at the nozzle exit, it is necessary to give the contour of the nozzle a particular shape ensuring a progressive supersonic expansion. Such a nozzle will be called contoured. Because of its accuracy and very low cost of calculation time, the method of characteristics is well suited for the determination of the flow in supersonic planar or axisymmetric nozzles equipping wind tunnels or space launcher's rocket engines. As mentioned above, a supersonic nozzle has three parts:

- a subsonic domain where the non-viscous part of the flow (outside the boundary layers) is governed by a system of elliptic differential equations,
- a transonic domain,
- and a supersonic domain where the system of equations is hyperbolic and which can be computed by the method of characteristics.

The method of characteristics is well documented in compressible aerodynamic textbooks to which the reader is referred. Since this method cannot be applied to the entire flow, the calculation procedure consists of adopting a particular treatment for the transonic domain which makes it possible to determine the initial part of the supersonic domain from which the method of characteristics is applied.

To achieve a uniform flow of Mach number M_E at the nozzle exit area A_E, the throat cross section A_C, is first calculated by the formula:

$$A_c = \frac{A_E}{\sum(M_E, \gamma)}$$

where $\sum(M, \gamma) = A/A_c$ is the isentropic relation of flow in CD nozzles given earlier. The throat height, h_c, (two-dimensional nozzle) or the radius of the throat, r_c, (axi-symmetric nozzle) is deduced and a local radius of curvature, $\Re$, for the throat region is chosen. The method of defining the nozzle contour according to a so-called inverse procedure is decomposed into 5 steps.

1. The flow in the throat region is calculated by analytical methods based on the potential flow equation or by numerical solution of the Euler equations.
2. In a second step, a Mach number distribution is given along the axis of the nozzle $M(x)$ providing a continuous transition between the transonic domain and the desired constant level downstream, as shown in Fig. 5.3.
3. The flow downstream of the transonic domain is calculated by progressing along left running characteristics starting from the axis while taking into account the imposed distribution $M(x)$ on the axis (see Fig. 5.4).
4. At the end of step 3, the wall of the nozzle $P(x)$ is defined, keeping in mind that it is a streamline.

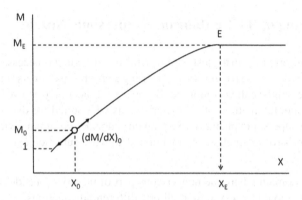

Fig. 5.3 Defining a supersonic nozzle. Distribution of the Mach number imposed on the nozzle axis

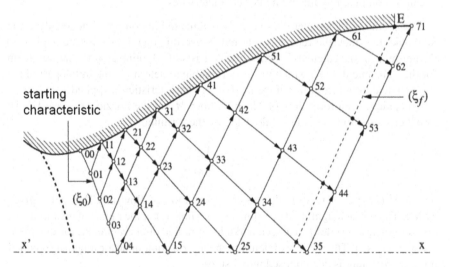

Fig. 5.4 Defining a supersonic nozzle by the method of characteristics marching scheme

5. To account for the development of the boundary layer along the nozzle wall, a boundary layer calculation is performed on $P(x)$, from which the evolution of the displacement thickness $\delta^*(x)$ is extracted. The final corrected wall, $P_{final}(x)$, is obtained by "thickening" $P(x)$ by adding the displacement thickness $\delta^*(x)$:

$$P_{final}(x) = P(x) + \delta^*(x)$$

The boundary layer correction is mandatory for the nozzles in hypersonic wind tunnels where, due to the high Mach number and the low density, the boundary layer thickens considerably, occupying a large part of the flow inside the nozzle. This correction is essential in so-called low-density installations, simulating flows at very high altitude.

Finally, the contour of the nozzle is defined by a circular arc in the throat region and a succession of points downstream. The choice of the ratio of the radius of curvature at the throat and the height (radius) of the throat is critical. A ratio, too small will lead to the formation of shock waves by focusing of characteristics. A value of $\Re/r_c = 4$ (or $\Re/h_c = 4$) is a minimum, ratios of at least 10 being adopted in nozzles for hypersonic wind tunnels.

In the following example, the method is applied to determine the contour of a planar two-dimensional nozzle to produce a uniform Mach number $M_o = 2.5$ in air ($\gamma = 1.4$). Figure 5.5 shows the Mach number distribution along the axis of symmetry of the nozzle. The inverse procedure leads to the contour shown in Fig. 5.6 which also represents the net of the calculated characteristics. Figure 5.7 shows the nozzle of a supersonic wind tunnel designed using the methods of characteristics.

The test-rhombus of a supersonic test section is the volume within which the supersonic flow is uniform and where the model to be tested can be mounted. It is delimited upstream by the characteristics leading to the nozzle extremity and

Fig. 5.5 Mach number distribution along the axis of a nozzle design for Mach 2.5

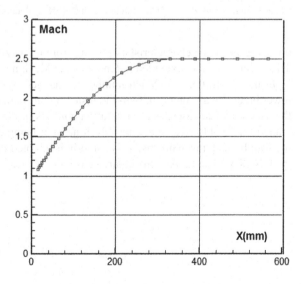

Fig. 5.6 Two-dimensional planar nozzle for a Mach 2.5 flow as a network of characteristics

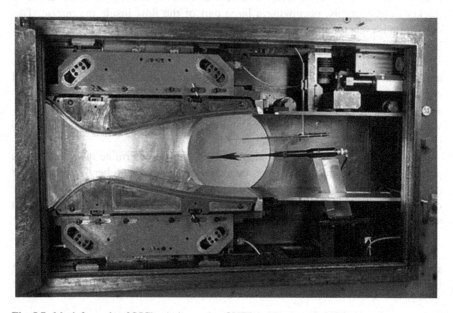

Fig. 5.7   Mach 2 nozzle of S5Ch wind tunnel at ONERA, Meudon (© ONERA)

downstream by the characteristics starting from the nozzle extremity (see Fig. 5.8). This volume is more extended in length as the Mach number is increased.

In most facilities, the Mach number is changed by replacing the upper and/or lower blocks (wall liners) in two-dimensional configurations or the entire nozzle in the axisymmetric cases. A rarer solution consists of designing a nozzle with walls made of a flexible steel sheet, the Mach number being changed by deformation of the wall by jack mechanisms. This solution is adopted in the S3MA wind tunnel at the ONERA Modane-Avrieux centre (see Sect. 5.3.4).

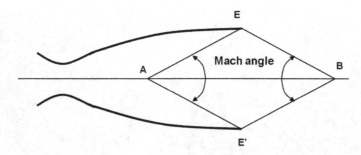

Fig. 5.8 Mach rhombus of a supersonic nozzle

5.3 Typical Supersonic Wind Tunnels

5.3.1 Basic Research Supersonic Wind Tunnel

The **S8Ch supersonic wind tunnel at ONERA, Meudon** is equipped with two test sections with identical characteristics and utilises the same power supply and common auxiliaries. Of an open circuit type, the wind tunnel operates continuously at atmospheric conditions at the intake and a downstream suction generated by two 132 kW pumps. Before expansion, the atmospheric air is dried by passing it over a bed of silica gel.

The facility has an auxiliary vacuum pump and a high-pressure air supply of 7 bar for the simulation of propulsion jets and to carry out flow control studies. The test sections have a square section of 0.12×0.12 m^2 in supersonic (up to Mach 2) and rectangular section, 0.12×0.10 m^2 in transonic regime. Total pressure and temperature are close to 10^5 Pa and 300 K respectively. Figure 5.9 shows one of the test sections of the wind tunnel.

These facilities are dedicated to studies on shock wave/boundary layer interactions and shock wave/turbulence interactions, separation control, cavity flows, jets, turbulence, laminarity, optical diagnostics, etc. A very complete set of measurement and visualisation techniques can be used for a fine characterisation of flows: various pressure probes, and optical based systems such as PSP, PIV, LDV, infrared thermography, schlieren and shadowgraph, surface flow visualisation. Figure 5.10 shows the test section in the half-nozzle configuration where the plane of symmetry of the nozzle is replaced by a solid flat wall. The rotating cam located in the second throat location makes it possible to periodically vary the position of the shock wave in the test section for unsteady analysis.

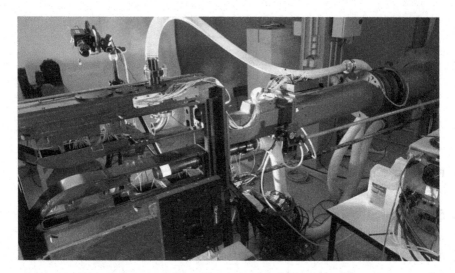

Fig. 5.9 One of the test sections of the ONERA S8Ch wind tunnel (© ONERA)

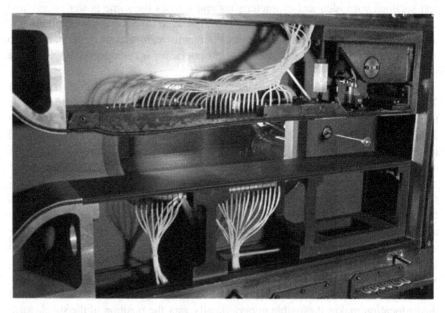

Fig. 5.10 Test section of the ONERA S8Ch wind tunnel. Half-nozzle assembly (© ONERA)

5.3.2 Low Turbulence Supersonic Wind Tunnel

The design of the **S8 research wind tunnel of IUSTI** (Institute of Industrial Thermal Systems at the University of Marseille) addresses several objectives. The aim here is to generate a freestream flow with very low background disturbances while reducing the effect of the tunnel wall turbulence as well. This is essential for very detailed analysis of the fluctuating components and turbulence structure in transitional and separating flows, using hot-wire anemometry, still the most convenient technique for the measurement of unsteadiness in flows. This involves constant monitoring of the quality of air, which has to be strictly dried and dust-free, and while maintaining constant ambient conditions. The minimum duration over which aerodynamic conditions must be kept constant depends on the characteristic time of the phenomena expected, the constraint being the time required for data acquisition to ensure an acceptable size of the sample, a duration that becomes important for conditional analyses. Hence the need of a wind tunnel operating continuously. Finally, a wind tunnel capable of operating at moderate Reynolds numbers is important for validation of numerical modelling such as LES or DES.

The main characteristics of the wind tunnel, which is equipped with two test sections, S7 and S8, are as follows (see Fig. 5.11).

– total pressure: 0.12 to 0.90 bar,
– total temperature: atmospheric,
– unit Reynolds number: 5.5×10^6/m at a total pressure of 0.5×10^5 Pa.

The wind tunnel is equipped with contoured liner nozzles of the following exit cross section:

– test section S7 ($M = 1.7$): 80×150 mm^2
– test section S8 ($M = 2.0$): 105×170 mm^2.

Fig. 5.11   S8 wind tunnel of the IUSTI at Marseille (© IUSTI)

The air supply and circulation is provided by a centrifugal compressor with three stages, driven by a motor of 450 kW. The cooling is through a heat exchanger operated at similar power to cool down the air to a temperature close to atmospheric. The settling section is equipped with 7 fine meshes to reduce the freestream turbulence prior to the beginning of the contraction. The tests sections are fixed to concrete blocks resting on vibration isolation or damping material, the various parts being connected to the rest of the wind tunnel by flexible joints in order to avoid the transmission of mechanical vibrations within the measurement zones. The air is desiccated by a molecular sieve dryer. The dust is removed by paper filter banks that extract 99% of particles larger than 1 μm. The pressure level in the circuit is regulated by a system maintaining a constant pressure with an accuracy of ±13 Pa, with a response time of a few seconds ensuring a very good aerodynamic stability.

The main devices for damping disturbances: In high velocity flow, the disturbances to be eliminated are mainly perturbations of temperature, pressure and freestream velocity or turbulence due to the mechanically power system. The temperature disturbances are produced by the compressor, where at the outlet it could reach about 200 °C. As mentioned above, the air is brought back to near atmospheric conditions by means of a large heat exchanger, which ensures unstratified flow in the ducts and in pressure gradient zones.

The pressure disturbances produced by the compressor are related to the rotation of the rotor which introduces a fluctuation of the fluid causing it to jolt at the outlet. These pulsations are eliminated by a Helmholtz filter tuned to the resonance frequency of the rotation speed of the compressor (similar to the anti-pogo device used on some rockets) and by organ pipes for the harmonics corresponding to the number of blades. Downstream of these devices, the discrete frequencies related to the mechanical system are no longer discernible on the spectra of the fluctuations. On the return part of the circuit, cavities prevent the noise produced mechanically from propagating towards the test sections from downstream of the nozzles.

The freestream turbulence intensity are reduced or damped by conventional means applied in the subsonic tunnel: careful design of corners with vanes to avoid the formation of longitudinal vortices; divergent portions with a small opening angle, possibly equipped with grids, so as to avoid separation; and quiet rooms as indicated above. The grid of the heat exchanger stages helps to destroy larger eddies. The very strict dust extraction from the air allows measurements by hot wire at a lower risk of damage. In the absence of natural contamination of the air, measurements by LDV or PIV require a seeding, the presence of the filter making it possible to control the particle size distribution. This seeding is recollected with a device downstream of the measurement zones.

Figure 5.12 shows a study of the reflection of a shock wave in the test section of the S7 wind tunnel. The shock wave is generated by a flat plate with sharp leading edge placed above the study wall.

Characteristics of the flow: The flows generated in the nozzles are largely free of upstream disturbances, the detected fluctuations being essentially the turbulence radiated by the boundary layers. For example, in a Mach 2.3 test section, with the boundary layer tripped by roughness on the side walls upstream of the throat and for

Fig. 5.12 Test setup for the study of the reflection of a shock wave in the test section S7 of IUSTI wind tunnel (© IUSTI)

a stagnation pressure of 0.5 bar, the fluctuation of (ρu) measured with the hot wire in a frequency band limited to 300 kHz is less than 0.1%.

At this Mach number, the threshold corresponds to a freestream turbulence intensity of less than 0.01% for both acoustic and vortical fluctuations. In comparison to the very low levels of turbulence of quiet tunnels (see Sect. 5.4), turbulence levels in the S8 test section are still remarkably low.

5.3.3 Transonic/Supersonic Open Jet Facilities

Installed on the **CEAT-PROMETEE test platform in Poitiers** (see Sect. 3.2.5), the compressible wind tunnels of the PPRIME Institute are dedicated to the study of transonic and supersonic flows. The facility consists of transonic/supersonic wind tunnels able to provide flows for duration of up to 10 min, powered by the 200 bar compressed air network of the platform.

– **The T200 wind tunnel** allows the study of coaxial open jets of up to 200 mm in diameter, the configuration consisting of a transonic to supersonic primary jet (maximum stagnation pressure of 150 bar), a secondary subsonic to transonic

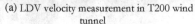

(a) LDV velocity measurement in T200 wind   (b) S150 wind tunnel in closed test section and
                    tunnel   free jet configuration

Fig. 5.13 T200 and S150 wind tunnels of CEAT-PROMETEE test platform in Poitiers (© Institut PPRIME)

jet (maximum stagnation pressure of 3 bar, Mach number from 0.5 to 1.3, see Fig. 5.13a).

- **The S150 wind tunnel** operates at a Mach number ranging from 0.8 to 2.8, with a total pressure of up to 40 bars, in a test of 150×150 mm^2 or in a free jet. It is used for shock wave/turbulence interaction studies or unsteady flows in launchers propulsive nozzles (see Fig. 5.13b).

- **The MARTEL bench** (see Fig. 5.14) is devoted to supersonic jets at high temperature in steady state mode (up to Mach 3, stagnation temperature of 1800 °C) or in transient mode generating a blast wave (Air-CH4 deflagration up to 200 bar and 2200 °C) in an open semi-anechoic test hall.

Instrumentations consist of acoustic and velocity (LDV, 2D and 3D PIV), unsteady pressure and density measurement techniques. Each bench allows continuous or instantaneous schlieren visualisations also. These means allow the characterisation, understanding and modelling of supersonic flows for various subjects such as basic aerodynamic instabilities, separated flows in nozzles, impinging jets, wall boundary layers, shock wave/turbulence interaction, noise generation mechanisms, compressibility effects on turbulence, etc.

5.3.4   Large Variable Mach Number Wind Tunnel

The **S3MA wind tunnel at the ONERA centre in Modane-Avrieux**, (see Fig. 5.15) is a transonic/supersonic wind tunnel that can deliver flows from subsonic to supersonic. The different nozzle configurations are shown in Fig. 5.16. In subsonic-transonic regimes, the wind tunnel is equipped with interchangeable test sections of 0.56 m width with perforated walls on all four sides. In supersonic regime, the wind tunnel is equipped with a nozzle of height 0.80 m and width 0.76 m.

The wind tunnel has a set of 5 supersonic nozzles whose nominal Mach number ranges from 1.5 to 5.5. Intermediate values between Mach 4.5 and 5.5 can be realised continuously by rotating the nozzle blocks so as to change the ratio of outlet area

Fig. 5.14 MARTEL bench of CEAT-PROMETEE test platform in Poitiers (© Institut PPRIME)

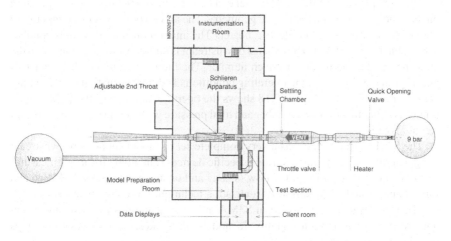

Fig. 5.15 Layout of the ONERA S3MA wind tunnel (© ONERA)

Test sections

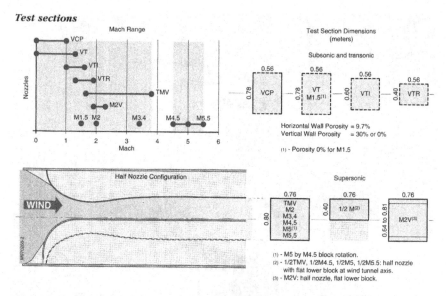

Fig. 5.16 Configurations of the nozzle of the ONERA S3MA wind tunnel (© ONERA)

to throat area. A nozzle with deformable walls allows a continuous variation of the Mach number from 1.65 to 3.8 (see Fig. 5.17). Two sources of compressed air are available; 9500 m^3 under a pressure of 9 bars and 115 m^3 under a pressure of 270 bars. A vacuum tank of 8000 m^3 at a minimum pressure of 0.010 bar is also available.

The total pressure of the wind tunnel can be varied from 0.2 to 7.5 bar depending on the Mach number and the nozzle geometry. The air can be heated to a temperature of 530 K by means of an electric heater. The maximum unit Reynolds number achievable is 54×10^6/m. The exhaust downstream is either to a vacuum tank or in the atmosphere, depending on the Mach number. The blow-down time is between 10 s and 15 min depending on the operating conditions and the evacuation mode (empty sphere or atmosphere). Figure 5.18 shows the operational range of the S3MA wind tunnel as a function of Mach number and total pressure with the curves of the constant Reynolds number.

The S3MA wind tunnel has a very complete set of model supports (downstream sting, lateral supports, masts, various specific devices) and many means of measurement and characterisation of the flow (steady and unsteady pressures, PSP, two and three-dimensional LDV, various visualisation techniques, 6-component balances).

The tests typically performed in S3MA are; steady and unsteady force/moment and pressures measurements on missile models, air intakes, optronic systems, wings or helicopter rotor profiles.

Fig. 5.17 Supersonic nozzle with a variable Mach number (1.65–3.8) of the ONERA S3MA wind tunnel (© ONERA)

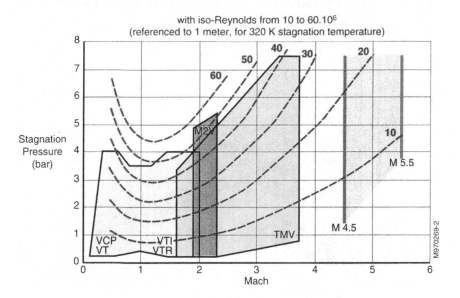

Fig. 5.18 Simulation domain of the ONERA S3MA wind tunnel (© ONERA)

5.3.5 Blow Down Transonic/Supersonic Wind Tunnel

The origin for the construction of the **transonic/supersonic Mach 3 wind tunnel of the ISL** (French-German Research Institute of St. Louis) in 2015 is the result of a reflection on future guided munitions. This research wind tunnel is used for the study of the aerodynamic behaviour of models driven by different types of actuators. The air is compressed by two compressor lines producing a mass flow of 1800 kg/h at 30 bar, stored in 9 tanks with a total capacity of about 288 m³. The air is expanded by a fast-response valve to obtain a total pressure ranging between 2 and 18 bar. The air flow is accelerated through a converging-diverging nozzle and then vented to the atmosphere through an exhaust equipped with a silencer, for Mach number ranging from 0.30 to 4.50 (see Fig. 5.19).

The maximum flow rate of the wind tunnel is 54 kg/s with a the blow-down duration of 30 to 120 s, depending on relation between Mach number and total pressure. A transonic test section is available for Mach numbers less than 1.20 and another supersonic test section is used for higher Mach numbers. The transonic test section has a second throat and is equipped with perforated walls with a secondary exhaust (see Fig. 5.20). The Mach number of the supersonic test section is adjusted by a nozzle with deformable walls so that the flow conditions can be changed during a single run.

The test section is 400 mm high and 300 mm wide and can accommodate models with a diameter at most equal to 40 mm and maximum length of 300 mm without choking the flow. These dimensions are sufficient to easily reproduce the details of full scale models, the test section allowing the study of models at incidence ranging

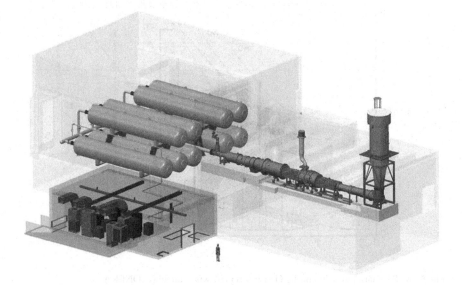

Fig. 5.19 General organisation of ISL trisonic wind tunnel (© ISL)

Fig. 5.20 Test section of the ISL trisonic wind tunnel (© ISL)

from $-7°$ to $+24°$. The Reynolds number, based on the maximum diameter of the model, is between 5×10^5 and 2.2×10^6, for a turbulence intensity of less than 1%.

5.4 Mach 6 Quiet Wind Tunnel

The challenges involved in reproducing the phenomenon of laminar to turbulent transition in a wind tunnel due to its sensitivity to ambient disturbances produced by the facility itself has already been pointed out in previous chapters. Figure 5.21 illustrates the sources of environmental disturbance in a supersonic or hypersonic wind

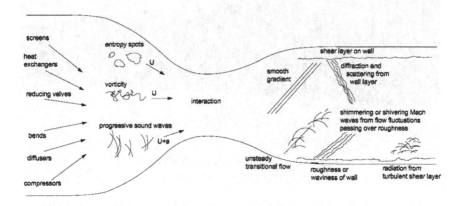

Fig. 5.21 Perturbations sources in a supersonic wind tunnel (© Purdue University)

tunnel. Upstream of the nozzle throat, the heaters, corners, diffusers, compressors, filters, are the source of disturbances in the form of entropy spots of thermal origin or freestream disturbance which constitutes of vortices and acoustic waves. Downstream of the throat, the nozzle's divergent wall is the main cause of disturbances due to surface non-uniformity generating Mach waves, roughness and turbulent boundary layer. A first condition to ensure a flow with low disturbances consists in generating upstream conditions as undisturbed as possible using so-called silent valves. The quiet wind tunnel of Purdue University uses a Ludwieg tube producing a very clean flow but for a short duration. An axisymmetric nozzle is also preferred to avoid instabilities related to perturbations produced by corners.

Thus, supersonic/hypersonic wind tunnels and shock tunnels (see Chap. 6) are affected by high levels of upstream flow fluctuations, typically one or two orders of magnitude higher than the real flight cases. These fluctuations are most often dominated by the noise radiated by the turbulent boundary layers developing on the wall of the nozzle whose turbulent structures radiate acoustic waves. Although often weak and negligible, this noise has a considerable effect on the laminar to turbulent transition on the models. It can also have a significant influence on other phenomena, such as separation. Quiet wind tunnels have been developed to minimise upstream disturbances and provide a laminar nozzle boundary layer to produce uniform flow at supersonic and hypersonic velocities with comparable noise levels encountered in real flight.

Figure 5.22 shows the noise radiated by a turbulent boundary layer from a magnified shadowgraph image obtained in the ballistic range of the Naval Ordnance Laboratory. The near zero incidence sharp nose-cone flies from the left to right in

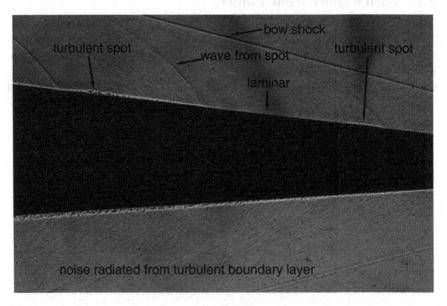

Fig. 5.22 Shadowgraph of the flow over a cone at Mach number 4.3 (© Purdue University)

static air at Mach 4.3 and a Reynolds number of 10×10^6/m. The photo reveals the acoustic waves radiated by the turbulent structures of the boundary layer on the lower surface of the cone. These waves propagate according to the Mach angles, defined from the flow velocity minus the velocity of the boundary-layer disturbances generating the acoustic waves.

On the upper surface, the boundary layer is intermittent with two clearly visible turbulent spots immersed in the laminar part. More pronounced waves are visible in front of the turbulent spots due to the increase in the displacement thickness, with lower levels of noise radiated by the turbulence inside the spots. Noise is not present above the laminar regions. This result shows that the control of the laminar to turbulent transition on the nozzle walls is essential to design facilities for the study of transition under conditions comparable to those in flight.

The Boeing/AFORS Mach 6 quiet wind tunnel was built at Purdue University to achieve quiet flows at low hypersonic Mach numbers (high supersonic!) and moderate Reynolds numbers in "cold" flow; i.e., without simulation of high-speed flow total enthalpy levels (see Sect. 6.2). Figure 5.23 shows the design of the Ludwieg tube installation consisting of a long tube terminated by a convergent-divergent nozzle; the flow enters the test section equipped with a second throat. In a Ludwieg tube, a diaphragm located downstream of the test section is abruptly burst this generating an expansion wave which propagates through the test section, the measurements are made in the gas downstream of the expansion wave. The low value of the Mach number in the driver tube allows the operation of the wind tunnel during many reflection cycles of the expansion wave, resulting in a test time of several seconds. This enables the generation a flow free of any disturbances due to pressure reducers and valves. In return, the test time is limited to a few tens of second.

To keep the boundary layer over the nozzle wall laminar, a suction slot at the throat of eliminates the convergent boundary layer, allowing a "fresh" boundary layer to develop in the diverging part. In addition, the wall of the throat and the diverging section is highly polished to remove the roughness and surface irregularities that would trigger transition. The very long nozzle has a large radius of curvature in the

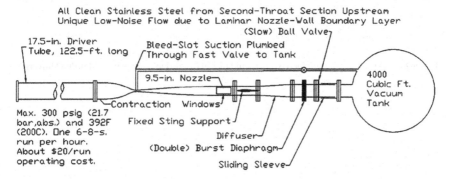

Fig. 5.23 Layout of the Mach 6 quiet wind tunnel of Purdue University (© Purdue University)

throat region in order to reduce the development of so-called Görtler instabilities along the wall of the divergent. The maximum permissible stagnation pressure for quiet operation is 11.7 bar. The wind tunnel can also operate with a noise level comparable to that of conventional wind tunnels without suction of the boundary layer at the throat.

Figure 5.24 shows a sketch of the flow at the nozzle outlet in the presence of a slender cone close to the maximum size allowing flow to start at zero-angle of incidence. The uniform flow begins at x = 1.91 m downstream of the throat and ends at x = 2.59 m, the evolutionary part of the nozzle ending at x = 2.58 m. The rectangular frame shows the location of observations 8 portholes. The Mach lines drawn in Fig. 5.24 are intended to show the origin of the radiated noise for different positions of transition on the nozzle wall. The boundary layer remains laminar downstream of the nozzle outlet for a total pressure greater than 11 bar, well beyond the initial design conditions.

The test section offers optical access through machined acrylic windows to conform to the nozzle contour. Most measurements use heat-sensitive paints to visualise the surface heat flux and high-frequency unsteady pressure transducers to detect unstable waves. Other means of measurement are also available.

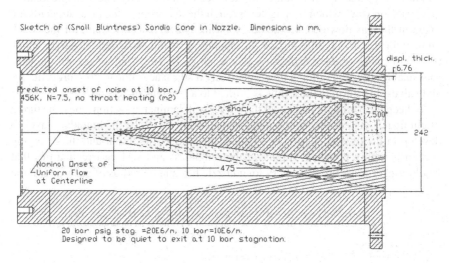

Fig. 5.24 Test section of the Mach 6 quiet wind tunnel of Purdue University (© Purdue University)

Chapter 6
Hypersonic Wind Tunnels

6.1 Types of Hypersonic Wind Tunnels

Hypersonic wind tunnels are intended for studying the aerodynamics of hypersonic aircraft, space launchers, missiles, projectiles and the atmospheric re-entry of space vehicles. In principle, there are no differences between the operation of a hypersonic and a supersonic wind tunnel, except that the contraction ratio of the nozzles, A/A_c reaches very high values for higher Mach number, for instance an $A/A_c = 586$ is needed to achieve a Mach number of 10. Hence throat dimensions are so small that two-dimensional nozzles are most often abandoned in favour of axisymmetric nozzles. In addition, establishment of the flow requires considerable upstream/downstream pressure ratios, approximately 3300 for Mach 10, obtained by compressing air upstream and expanding it downstream while filling a vacuum tank. The energy required to run such a facility makes it such that the hypersonic wind tunnels are most often of the blow down type. The energy required for a test is usually stored in the form of a compressed gas or a flywheel rotating at high speed which is a time-consuming process, but then released for a very short time during the experiment. As a result, the duration of a test is limited, from a few seconds and could reach even several minutes, for "cold" installations, but to a hundred milliseconds for "hot" installations.

At high Mach number adiabatic expansion in the nozzle is accompanied by an intense cooling necessitating heating of the air to prevent liquefaction. Thus, the minimum level of upstream temperature allowed at Mach 10 is 1100 K (~830 °C), which leads to a temperature in the test section of 52 K (~ −220 °C), at the limit of the liquefaction of oxygen. Such facilities, where the gas is heated just enough to prevent its liquefaction during the expansion, are called "cold" hypersonic wind tunnels. In such wind tunnels, the duration of the test ranges from a few seconds to several minutes depending on the conditions of the run and the characteristics of the components of the facility, including the diffuser.

© Springer Nature Switzerland AG 2020
B. Chanetz et al., *Experimental Aerodynamics*,
Springer Tracts in Mechanical Engineering,
https://doi.org/10.1007/978-3-030-35562-3_6

High Mach numbers are not the only specification of hypersonic flows encountered in space applications. Indeed, when a vehicle enters the atmosphere at a speed of several kilometres per second, the friction from flow on the body leads to temperatures of several thousand degrees (adiabatic compression). This results in considerable wall heat fluxes and the triggering of chemical reactions within the air due to non-equilibrium between internal energies, dissociation of molecules and chemical reactions between molecules. The temperature is a measure of the energies of the translational, vibrational and rotational motions of the constituents of a gas: molecules, atoms and subatomic particles. In the so-called equilibrium state, energy is distributed evenly among these various motions by energy exchanges. When the gas is subjected to a rapid variation in conditions (expansion in a hypersonic nozzle, or compression by a shock wave) the gas adjusts more or less quickly to the new conditions: the translation occurs almost instantaneously, as well as the vibration but the rotation being much slower. The gas is then said to be in non-equilibrium and no longer corresponds to the model of the perfect gas: these effects are the so-called real gas effects whose behaviour deviates from that of the perfect gas. A true hypersonic wind tunnel must therefore reproduce not only large Mach numbers, but also very high temperature levels; it is then called a hot hypersonic or hyper-enthalpic wind tunnel.

6.2   Hypersonic "Cold" Wind Tunnels

Although the actual phenomena of hypersonic flight are not fully simulated, these wind tunnels are a valuable tool for studying the aerodynamics at high Mach numbers. In the most common case, and for generating stagnation temperatures not exceeding 1000 K, the heating system is made of a heater and a heat exchanger made of metal plates or spheres. Before the test, the exchanger is heated by circulation of electrically heated air, this rather long phase requiring only moderate power. In another technique, the flow passes through a bed of alumina beads heated by a secondary stream of high temperature air generated by the combustion of propane. In an instantaneous mode of heating, the test gas passes into tubes heated by Joule effect, at the expense of higher electrical power. The advantage of the latter method lies in attaining temperatures of up to about 1200 K and the absence of dust generated in heat exchangers.

These facilities are of the long run type (from 10 s to several minutes) operating on a reservoir of compressed air. **The R1Ch, R2Ch and R3Ch wind tunnels at ONERA Meudon centre**, for instance operate on a reservoir at 250 bar and a vacuum sphere of 500 m^3 shared between the three facilities. For the R1Ch and R2Ch wind tunnels, before expansion the air in the nozzle is heated by passing through a heat exchanger that can attain a temperature of 700 K at a flow rate of 40 kg/s. An auxiliary high-pressure air supply (200 bar) is used to simulate propulsive jets. R1Ch and R2Ch wind tunnels cover the range of supersonic to hypersonic Mach numbers, between 3 and 7 and unit Reynolds numbers ranging between 2.1×10^6/m and 5.1×10^6/m using variations in the stagnation pressure from 0.5 to 80 bar. These wind tunnels use

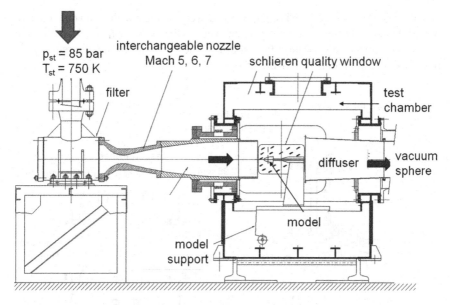

Fig. 6.1 The R2Ch wind tunnel at ONERA, Meudon (© ONERA)

a set of 70 compressed air storage cylinders at 250 bar, representing a total volume of 30.5 m³. Figure 6.1 shows the layout of the R2Ch wind tunnel.

The R1Ch wind tunnel can be equipped with Mach 3 or 5 contoured nozzles with an exit diameter of 0.326 m, the maximum stagnation pressure and temperature being 15 bar and 400 K respectively. The R2Ch wind tunnel can be equipped with Mach 3 and 4 nozzles with exit diameter of 0.190 m, or Mach 5, 6 and 7 nozzles with exit diameter of 0.326 m, with a maximum total pressure and temperature of 80 bar and 700 K respectively. Depending on the test configuration (model size, stagnation pressure, Mach number), the flow is discharged either to the atmosphere at ambient pressure, in which case the run can last several minutes, or in the vacuum sphere of 500 m³ where the pressure can be as low as a few millibars. In the latter mode, the effective run time is about thirty seconds. The duration of the test is in fact controlled by the pressure downstream of the test section, which increases with the level attained in the discharge sphere until reaching a level incompatible with the maintenance of the hypersonic flow. The diffuser is an essential component for this type of wind tunnel, especially those called low density facilities simulating conditions at very high altitude (pressure in the test section of a few Pascals). The recovering ability of the diffuser, i.e., its ability to compress the flow that has been expanded in the test section, directly affects the duration of the run. A high pressure at the exit of the diffuser increases the pressure in the sphere before the flow un-starting, hence a longer run time.

In order to achieve a Mach number of 10, the R3Ch wind tunnel is equipped with a Joule instant heater which can raise the temperature of the air to 1100 K (see Figs. 6.2

and 6.3). The free test section is equipped with a contoured nozzle of an exit diameter
of 0.350 m and the stagnation pressure can vary in the range of 12–120 bar. A three-
way valve achieves a very short flow starting time (a few milliseconds) required for
heat flow measurements (see Sect. 9.5).

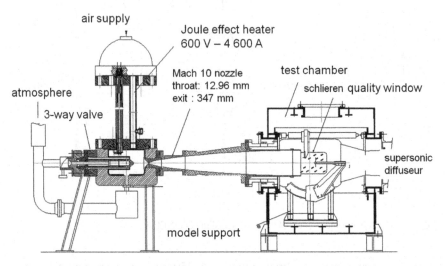

Fig. 6.2 The R3Ch wind tunnel at ONERA, Meudon (© ONERA)

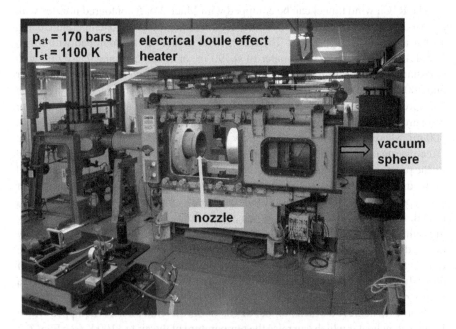

Fig. 6.3 View of the R3Ch wind tunnel at ONERA, Meudon (© ONERA)

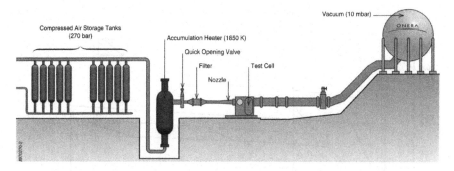

Fig. 6.4 Layout of the S4MA wind tunnel at ONERA, Modane-Avrieux (© ONERA)

A range of measurement can be performed in these wind tunnels, such as: force balances, steady and unsteady pressure, LDV, PIV, PSP, infrared thermography, Schlieren visualisation and heat transfer.

Another hypersonic facility of the same family is the **S4MA wind tunnel at ONERA Modane-Avrieux centre** (see Fig. 6.4), it produces a flow at a maximum stagnation pressure of 120 bar and can attain a maximum stagnation temperature of 1800 K, higher than those mentioned above.

It is equipped with axisymmetric nozzles with an exit diameter of 0.685 m, for Mach 6 and 4, and 0.994 m for Mach 10 and 12. The air passes through a heat exchanger containing 10 tons of alumina spheres heated by propane combustion before the test. The upstream dry air is stored at a pressure of 270 bar in tanks with a total capacity of 29 m^3 and the downstream vacuum sphere (minimum pressure of 0.01 bar) has a volume of 8000 m^3. For heat transfer measurements, the model is immersed into the flow by a fast-motorised support system.

The test procedure is as follows: once the temperature of the alumina spheres is established, the tank that stores the heated air is slowly pressurised to the desired total pressure. Then the rapid response actuator valve is open through an automated system for stabilisation and control of the stagnation conditions. Upon establishing the desired flow speed the model is introduce and, followed by acquisition of data, the model is then ejected out of the flow and the heater is purged at the end of the run. The useful test time, of 25–85 s, varies with the Mach number and the chosen stagnation conditions. Figure 6.5 shows the operating envelope of the wind tunnel based on stagnation conditions.

The S4MA wind tunnel is equipped with similar instrumentation to that of the R1Ch, R2Ch and R3Ch wind tunnels. Typical tests include force measurements on a complete model as well as on model parts (hinge moment of control surfaces). Figure 6.6 shows a space vehicle model in the test section of S4MA.

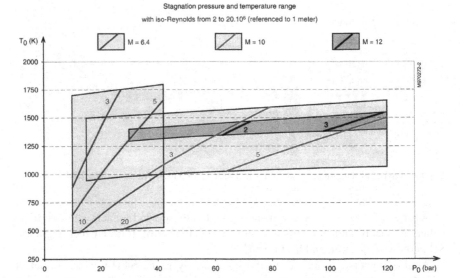

Fig. 6.5 Operation range of the ONERA S4MA wind tunnel (© ONERA)

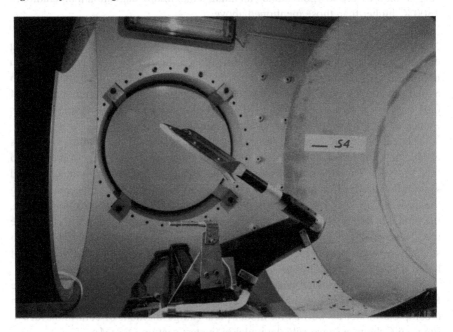

Fig. 6.6 Space vehicle model in the ONERA S4MA wind tunnel (© ONERA)

6.3   Hypersonic "Hot" or Hyper-enthalpic Wind Tunnels

To attain stagnation temperatures higher than 2500 K, with pressures that can exceed 1000 bar, considerable power is required. The duration of a run in this type of wind tunnel is most often extremely short, also limited at high temperature by the mechanical strength of the materials constituting the upstream part of the nozzle. Then hot shot wind tunnels type solutions are considered. In this category also enters the plasma wind tunnels using a gas ionised by the passage of an electric current.

6.3.1   Hot Shot Wind Tunnels

In a hot shot wind tunnel, the energy is provided by an arc chamber filled with the fluid to be heated (air or nitrogen) at a temperature and a pressure chosen according to the desired final conditions. The F4 high enthalpy wind tunnel at ONERA, Fauga-Mauzac was designed to study the re-entry phase of space vehicles with real gas effects. The test gas (air, nitrogen or CO_2), contained in an adjustable volume chamber of 1–14.7 litres, is brought to the test conditions by a high intensity electric arc, springing between an electrode and the wall of the arc chamber (see Figs. 6.7 and 6.8). The current is provided by a generator with a power of 150 MW consisting of a 15 tonnes flywheel rotating at 600 rpm, which represents energy storage of 400 MJ. The bursting of a diaphragm located at the nozzle throat releases the flow in the nozzle where Mach numbers up to 20 are obtained, for run duration of 200 ms. A pressure

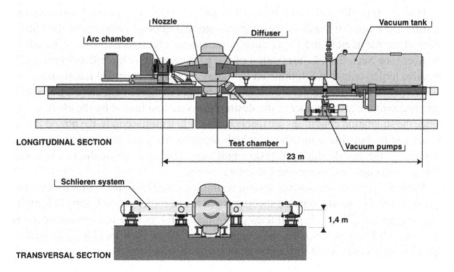

Fig. 6.7 Layout of the circuit of the F4 wind tunnel at the ONERA Fauga-Mauzac centre (© ONERA)

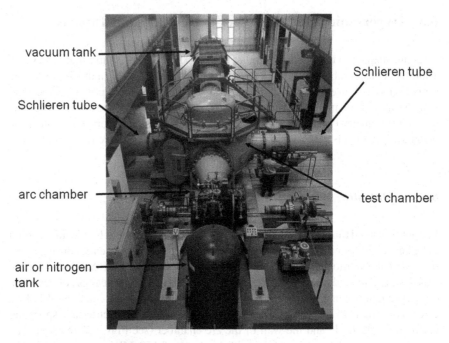

vacuum tank

Schlieren tube

Schlieren tube

arc chamber

test chamber

air or nitrogen
tank

Fig. 6.8 View of the ONERA F4 wind tunnel (© ONERA)

of 1000 bar and a specific enthalpy of 16.5 MJ/kg for temperatures of 3000–8000 °C
can be attained.

Four contoured nozzles of 0.670, 0.430 and 0.930 m exit diameter allow testing
at different ranges of the dissociation parameter based on ρL, also called the binary
interaction parameter, and of rarefaction parameter characterised by the Knudsen
number (see Sect. 2.4). Conservation of the parameter ρL is intended to reproduce
reactive hypersonic flows dominated by binary molecular chemical reactions.

To operate the wind tunnel the electrical generator is started first, the test gas is then
pumped into the arc chamber at the desired pressure and heated by the electric arc.
Once the stagnation conditions are reached, after a few milliseconds, the nozzle throat
is opened by bursting a membrane which separates the upstream and downstream
regions of the nozzle during pressure build-up. The duration of the run is about
200 ms and ended by triggering a discharge valve.

Various means of measurements can be implemented similar to other hypersonic
wind tunnels but more complex spectroscopic measurements (see Chap. 12) can be
also implemented here. Typical tests performed in F4 include force measurements
on space glider models and re-entry capsules at high angle of incidence, as well as
measurements of thermal flux and wall pressure distributions.

6.3.2 Shock Tubes and Shock Tunnels

The shock tunnels are well adapted facilities for the study of objects under conditions of hypersonic flight since conditions corresponding to the atmosphere of different planets can be reproduced. In addition, the tests are performed at low cost compared to those in wind tunnels equipped with a heater powerful enough to achieve the right temperature in the flow. The shock tube and its variants, such as the shock tunnel, are simpler means to achieve supersonic/hypersonic flow with high levels of pressure and total enthalpy, while the energy spent to operate a shock tube being modest. In return, the durations of a useful flow are very brief (from a few hundred microseconds to a few milliseconds). The shock tube is also used to study the conditions of formation of shock waves (detonation), the phenomena resulting from the propagation and reflection of shocks and also for the analysis of rapid processes involved in chemical kinetics or nuclear reactions.

A shock tube consists of a cylindrical tube of circular or square cross section, closed at its ends and divided into two initially isolated compartments (see Fig. 6.9). The low pressure compartment contains the test gas, at a desired pressure and temperature, and the high-pressure compartment filled with the driver gas at high pressure of several hundreds or even thousands of bars and at temperature similar to that of the test gas. The low-pressure section is 5–10 times longer than the high-pressure section. The driver gas and the working gas are generally of different natures and separated by a diaphragm. The instantaneous bursting of the diaphragm puts the two gases in contact which are then separated by an interface which cannot be maintained in the equilibrium state. The interface then propagates towards the low pressure like a piston generating (see Fig. 6.9).

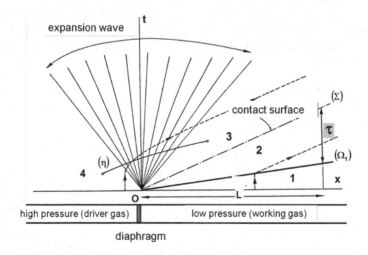

Fig. 6.9 Operation of a shock tube and the propagation of the waves

– a shock wave in the working gas, propagating towards the right-hand side which
  passes from state 1 to state 2 characterised by a temperature which can be very
  high;
– a centred expansion wave in the driver gas, propagating towards the left-hand side
  forces the flow passing from state (4) to state (3); the fluids in states (2) and (3)
  are separated by the contact surface, also called interface.

The intensity of the shock produced is higher as the pressure p_4 is higher compared
to p_1 and the density of the driver gas is lower, which favours the use of helium or
hydrogen, which can be heated also. The shock tube is a simple means of obtaining
a flow at high pressure and temperature which may exceed 8000 K, for tubes using
hydrogen as driver gas.

In shock tubes, or shock tunnels, the working gas is used to feed a nozzle with a
large expansion ratio so as to achieve both a hypersonic (high Mach number) and a
hyper enthalpic (high total enthalpy) flow. After the passage of the shock, the nozzle
placed at the end of the tube is fed for a very short time (a few milliseconds) by the
working gas in the state 2. The level of stagnation enthalpy can be increased by considering a reflected shock as shown in Fig. 6.10, the working gas being compressed
again by the passage of the shock which is reflected at the back of the tube. As shown
in Fig. 6.11, a second diaphragm D_2 is then present at the end of the tube, upstream
of the nozzle. After reflection of the shock on the diaphragm, D_2, the gas is in state 5
where pressure and temperature are even higher. After the reflection, the diaphragm
D_2 breaks and the gas in state 5 propagates through the nozzle. A tailoring of the
shock wave interface makes it possible to optimise the run time.

The two independent high enthalpy facilities **STA and STB of the high-speed
flow laboratory of ISL** (French-German Institute of Saint-Louis) are able to provide
8 MJ/kg to perform tests in high-speed flows (see Figs. 6.12 and 6.13).

The conditions at sea-level and up to 70 km altitude can be reproduced for the flow
Mach number shown in Fig. 6.14. Contoured nozzles are available for Mach 3, 4.5,
6, 8 and 10, and conical nozzles for 3.5, 10, 12 and 14. The high-pressure tubes of

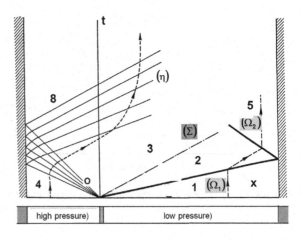

Fig. 6.10 Shock tube with
reflected shock effect

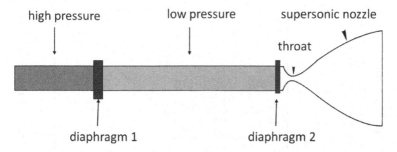

Fig. 6.11 Sketch of a reflected shock tunnel

Fig. 6.12 STA shock tunnel of ISL (© ISL)

Fig. 6.13 STB shock tunnel of ISL (© ISL)

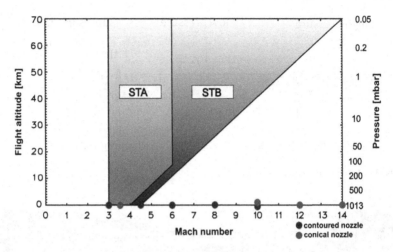

Fig. 6.14 Operating range of ISL shock tunnels STA (yellow) and STB (red) (© ISL)

STA and STB, with an interior diameter of 0.100 m are 3.6 and 4 m long respectively, and the low-pressure tubes are 18 m for both installations. The low-pressure tube is separated from the nozzle by a Mylar membrane; the first metal membrane is machined very precisely to withstand the pressure difference between the two tubes before bursting.

The high-pressure tube is filled with a lighter gas, typically hydrogen, at a pressure below 450 bar. The metal diaphragm separating the high and low-pressure compartments is made to burst at the pressure chosen according to the experimental conditions to be reproduced. Upon rupture of the diaphragm, the shock wave propagating in the low-pressure tube compresses, heats and accelerates the test gas to the end of the low-pressure tube where the Mylar membrane is placed. The reflection of the incident shock wave at the end of the low-pressure tube brings the test gas to rest for a very short time before the Mylar membrane gives way to quasi-stationary conditions, forming the stagnation conditions of the flow. The test gas then expands in the nozzle to generate a quasi-steady supersonic or hypersonic flow in the test section containing the model to be studied. This model can be supported by a sting or suspended by wires that break upon impact with the test gas. This free flight technique is interesting for the analysis of the dynamic behaviour of a model and the determination of some aerodynamic coefficients taking into account the base flow. At the end of the test, the test section and the gas recovery tank collect the working gas and the driver gas before being evacuated. These tanks have a volume of 10 m³ and 20 m³ for the STA and STB tubes respectively.

After each run, the flow conditions are calculated using a one-dimensional code reproducing the operation of the shock tunnel. This code requires the propagation velocity of the shock wave in the low-pressure tube, which is measured by a series of pressure sensors installed along the tube. One must also know the stagnation pressure which is measured by a Pitot tube placed in the Mach rhombus of the flow.

6.3.3 High Enthalpy Shock Tunnel

The **High Enthalpy shock tunnel in Göttingen (HEG)** is part of the German Aerospace Centre (DLR) and one of the major European hypersonic test facilities. It was commissioned in 1991 and has been utilised extensively since then in a large number of national and international space and hypersonic flight projects. Originally, the facility was designed for the investigation of the influence of high temperature effects such as chemical and thermal relaxation on the external surface of re-entry space vehicles. Over the last few years its range of operating conditions was subsequently extended. In this framework the main emphasis was to generate test section conditions which allow investigating the flow past hypersonic flight configuration from Mach 6 at low altitude up to Mach 10 at approximately 33 km altitude. The studies performed in HEG focused on the external as well as internal aerodynamics including combustion of hydrogen in complete supersonic combustion and the investigation of transition of the boundary layer from laminar to turbulent in hypersonic flow while testing transition delay techniques.

The HEG facility is a free piston shock tunnel in which the driver gas of the conventional shock tube is compressed by a piston in order to heat the driver gas before it comes into contact with the test gas, which increases the intensity of the shock. The operating principle is shown in the operating diagram in Fig. 6.15. The facility comprises a secondary reservoir, a compression tube, separated from the actual shock tube by a primary diaphragm, followed by a test nozzle and a recovery

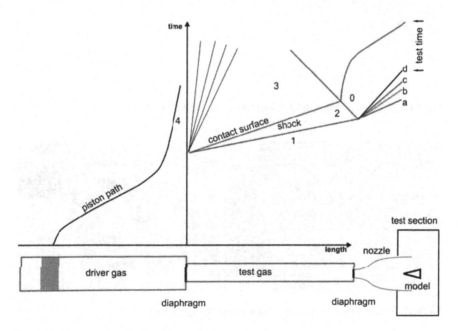

Fig. 6.15 Flow diagram of a free piston shock tunnel

tank. High pressure air, stored in the secondary tank, is used to accelerate the heavy piston along the compression tube. During this compression and the quasi-adiabatic heating of the light driver gas (helium or argon-helium mixture), the piston reaches a maximum speed of 300 m/s, the temperature of the driver gas increasing with the volumetric compression ratio. When the primary diaphragm bursts, a wave system similar to that which occurs in a conventional shock tube is formed.

The overall length of the HEG is 62 m and it weighs approximately 280 tonnes (see Figs. 6.16 and 6.17). A third of the weight is to reduce the tunnel recoil motion during

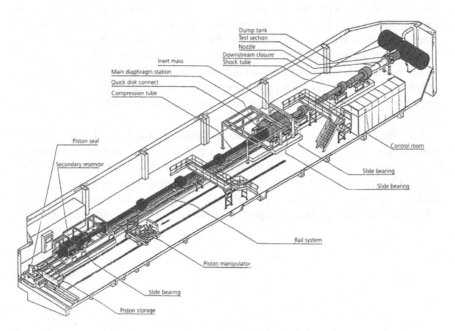

Fig. 6.16 The high-enthalpy shock tunnel HEG from DLR (© DLR)

Fig. 6.17 Views of the high-enthalpy shock tunnel HEG (© DLR)

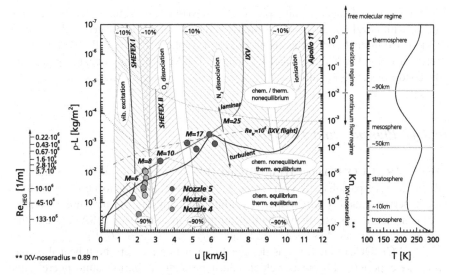

Fig. 6.18 Operating conditions of the HEG Göttingen shock tunnel (© DLR)

start-up. HEG was designed to provide a pulse of gas to a hypersonic convergent-divergent nozzle at total pressures of up to 200 MPa, and total enthalpies of up to 23 MJ/kg. Regarding the test gas there is no basic limitations. The operating conditions presented here are related to air as the test gas, but other operating conditions using nitrogen and carbon dioxide are also available. In order to correctly simulate the chemical dissociation occurring downstream of the bow shock of a re-entry vehicle the flight scaling parameter must be reproduced during ground-based testing. Further, the flow velocity is an additional driving parameter to be reproduced for high enthalpy testing. The operating conditions of HEG are shown in Fig. 6.18 in terms of the scaling parameter ρL and the flow velocity, u.

An indication of the corresponding flight altitudes is given on the right-hand side of Fig. 6.18, together with the temperature variation of the Earth's atmosphere. The Knudsen number indicates that the HEG operating conditions are within the continuum flow regime. Along a re-entry trajectory, the Reynolds number varies over several orders of magnitude, where in high altitude flight the wall boundary layer of a re-entry vehicle is initially laminar. Beyond a critical Reynolds number (shown as the curve labelled IXV in Fig. 6.18) the transition from a laminar to a turbulent boundary layer takes place. This process creates an increase in the skin friction and the wall heat flux. The HEG operating conditions (depicted with nozzle 5) are the original high enthalpy conditions covering a total specific enthalpy range from 12 to 23 MJ/kg.

Over the last few years the HEG operating range was subsequently extended. In this framework the main emphasis was to generate test section conditions which allow investigating the flow at hypersonic flight regime from Mach 6 at low altitude up to Mach 10 at approximately 33 km altitude. These low enthalpy conditions cover the range of total specific enthalpies of 1.5–6 MJ/kg. For a full scale wind tunnel model with nozzle 3, the M = 7.4 flight conditions can be replicated at 28 km and 33 km. Additional conditions using nozzle 4, replicates M = 6 flight conditions between 15 km flight altitude and at sea level. The M = 10 using nozzle 5 generates the conditions at 33 km altitude.

6.3.4 Plasma Wind Tunnels

The **SCIROCCO Plasma Wind Tunnel (PWT)** located at the Centro Italiano Ricerche Aerospaziali (CIRA) in Capua, Italy is the largest and most power consuming hypersonic facility which operates at high enthalpy and low pressure generated by an arc-jet. It was established by the European Space Agency (ESA) through the Hermes program and has been operational since 2002 (see Fig. 6.19).

The SCIROCCO facility was designed to generate the very high heat flux and pressure experienced by space vehicles during re-entry in Earth's atmosphere. This facility was designed to characterise real scale thermal protection systems (TPS), hot structures and payloads of space vehicles using the arc-jet technology.

It is operated by a 70 MW arc heater, capable of producing a 2 m diameter plasma jet at Mach 12, lasting up to 30 min. The arc heater (or plasmatron) is of segmented type and has a bore diameter of 0.11 m and a length of 5.5 m. The process air is

Fig. 6.19 Aerial view of the SCIROCCO plasma wind tunnel of CIRA (© CIRA)

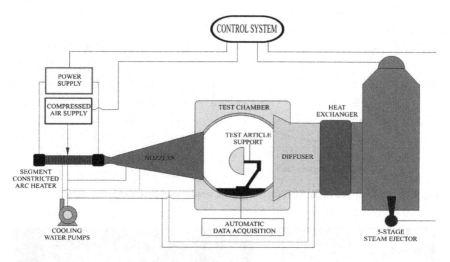

Fig. 6.20 Layout of the SCIROCCO plasma wind tunnel (© CIRA)

supplied to the arc heater at a pressure of 87 bar and at a mass flow rate of 0.2–3.5 kg/s, where it is heated up to plasma temperatures in the range of 2000–10,000 K.

Hypersonic speed is attained by accelerating the plasma flow through a converging-diverging conical nozzle, with exit diameters of 0.187–1.95 m. The five nozzle configurations offer the ability to test under a decent range of flow conditions.

Using an automated support system the experimental model or test article is inserted in the plasma jet which is confined inside the cylindrical test chamber, with an overall height of 9 m and an inner diameter of 5 m (Fig. 6.20).

The hypersonic jet is discharged into a 50 m long diffuser where the hypersonic to subsonic deceleration takes place and then is cooled by a powerful heat exchanger. A vacuum pump generates the desired low pressure conditions in the upstream test leg. The process gas is treated through the 'DeNOx System' before being released into the atmosphere, so as to remove the Nitrogen Oxides produced in the hypersonic-subsonic transition.

The operational envelope of the SCIROCCO plasma wind tunnel is shown in Fig. 6.21 in terms of total enthalpy and pressure on the right-hand side, and on the left-hand side in terms of simulated altitude and velocity compared to typical space shuttle re-entry trajectory.

The SCIROCCO facility is equipped with an extensive set of instrumentation in order to fully characterise the hypersonic jet flow conditions and its impact on the experimental model using:

– hot wall thermocouples, IR pyrometry and IR thermography (see Fig. 6.22),
– pressure sensors: Pitot/static,
– cold wall heat flux calorimetric sensors,
– optical flow diagnostics through emission spectroscopy (OES) and laser induced fluorescence (LIF).

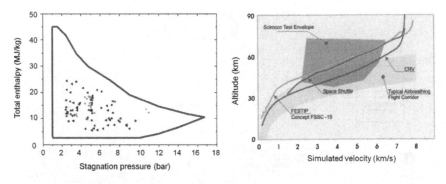

Fig. 6.21 Performance of the SCIROCCO plasma wind tunnel (© CIRA)

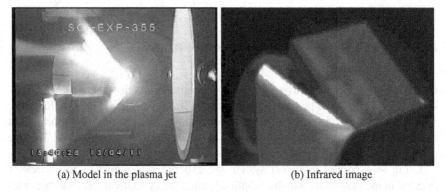

(a) Model in the plasma jet   (b) Infrared image

Fig. 6.22 Test on part of the ESA EXPERT vehicle in the SCIROCCO plasma wind tunnel (© CIRA)

The **GHIBLI** is another hypersonic, high enthalpy, low pressure arc-jet facility operated by CIRA for experiments on models of about 80 mm in diameter. It is mainly used for the characterisation and selection of shielding materials and also for CFD validations, for example on the stagnation point heat flux, hypersonic flow in intakes and nozzles, viscous interactions problems and shock–shock interactions. It is also useful for the characterisation of highly complex experimental measurement techniques required to capture aero-thermodynamic phenomena and derive empirical or semi-empirical correlations. The exploration of new measurement techniques is usually for heat flux or other flow parameters such as density, specific heat ratio, Mach number where non-intrusive methods are essential. The hot flow is generated by a 2 MW arc heater where the plasma flow is still at subsonic state and is later accelerated to hypersonic through a converging-diverging nozzle.

The nozzle exit section is 0.15 m in diameter and the plasma flow is accelerated to about Mach 10, depending on the boundary layer thickness.

The GHIBLI test chamber is a cylindrical compartment of 1.8 m inner diameter and 2 m length, where the plasma jet is fired normal to the longitudinal axis of the

Fig. 6.23 Arc heater and test chamber of the GHIBLI wind tunnel at CIRA (© CIRA)

chamber (see Fig. 6.23). After the interaction between the flow and the model surface, the plasma jet propagates to the diffuser and released to the atmosphere similarly to that of the SIROCCO plasma wind tunnel mentioned above. The test chamber walls are not cooled and have optical accesses for IR pyrometry and thermography, high speed cameras, spectroscopy and laser beams. Ports are distributed along the test chamber walls for the passage of pipes for pressure measurements or electric cabling to other devices. However, non-intrusive measurement techniques are also employed to characterise the free jet (see Fig. 6.24).

6.3.5 Continuous Plasma Wind Tunnel

The **Atmospheric Off-equilibrium Plasma Wind Tunnel (PHEDRA)** test facility of the ICARE Institute (Reactive and Environmental Aero-thermal Combustion Institute at Orléans) is another plasma wind tunnel equipped with a jet-type generator. It is a low-pressure facility (3–10 Pa) which allows a plasma flow of several kilometres per second. The advantage of this wind tunnel is that it generates a continuous and steady flow for several hours. The flexibility of its operation also makes it possible to simulate atmospheric re-entries on Earth or Mars. The research carried out on

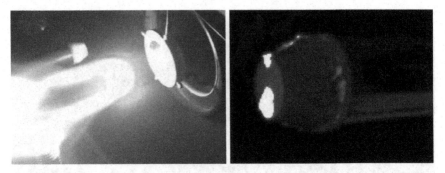

Fig. 6.24 Test of an aerothermodynamic configuration for space transportation in the GHIBLI wind tunnel (© CIRA)

these plasmas is based on a set of diagnostic means specifically developed to analyse the physicochemical properties of these media: electrostatic probes to determine the electron density, emission spectroscopy to study the molecules responsible for radiative flux, laser induced fluorescence to determine the velocity of plasma jets. Diagnostic means such as pressure and temperature probes and flow meters are also used, but not limited to these.

The PHEDRA wind tunnel consists of an experimental test section, a pumping unit, a plasma generator and a stabilised DC power supply. The test section is a horizontal steel cylinder 1.2 m in diameter and 3.2 m in length, with eight portholes in Plexiglas or aluminium of 0.5 m in diameter (see Fig. 6.25). An optical quality, high-strength glass porthole is used for optical measurements. Quartz or fluorine optical quality portholes can be fitted to the opposite windows to perform optical measurements with different instrumentations. A recirculating water system cools the back of the chamber (pumping side) which is directly exposed to the plasma jet.

The pumping unit consists of three vacuum lines arranged in series with a total capacity of $26{,}000\,m^3$/h. A pipe 20 m long and 0.4 m in diameter connects the bottom of the test chamber to the pumping unit via a valve 0.4 m in diameter. The plasma generator used in PHEDRA has been developed for the study of space probes atmospheric re-entry. The criteria guiding this design were: the use of gases composing of oxygen, high specific enthalpies with a very low cathode erosion rate, a very long operating time (several hours) at stable operating conditions, non-destructive starting of the electrodes and finally a design allowing easy maintenance.

The tunnel is operated by generating an electric arc of controlled intensity between the cathode and the throat of the nozzle used as the anode. The gas introduced into the convergent section of the conical nozzle is ionised at the passage of the throat and then accelerated in the divergent before expanding in the test chamber maintained at low pressure. A part of the energy of the arc transferred to the gas, flowing between the electrodes, heats, ionises and dissociates it. The other part is transferred to the cooling water circulating in the nozzle and the support of the cathode. The gas distribution has four independent inlets to simulate the composition of the atmosphere of different planets, such as Earth, Mars (97% CO_{2}-3 % N_2), Titan (99% N_2 − 1% CH_4) or Venus

Fig. 6.25 The PHEDRA wind tunnel of the ICARE institute (© Institut ICARE)

(96.5% CO_{2-3}.5% N_2). The low gas flow rates necessary for the operation of this generator allows for specific enthalpies high enough to simulate certain radiative properties of the atmospheric re-entry conditions.

As an example, Fig. 6.26 shows the study on the interaction of supersonic plasma of air at the stagnation point of a disk, at a pressure of 2.3×10^{-2} mbar.

More fundamental studies are also carried out to observe the effects of magneto-hydro- dynamics (MHD) on ionised flows in order to modify the shock wave around an obstacle. Figure 6.27 shows a supersonic flow of argon interacting with a hollow truncated cylinder, in which permanent magnets are inserted to create a secondary magnetic axial flow field.

The high altitude supersonic and hypersonic flight conditions can be simulated in the **MARPHy supersonic/hypersonic wind tunnel** at the ICARE institute in Orléans, whose operating conditions are shown in Fig. 6.28. The characteristics of the MARPHy wind tunnel makes it well suited for fundamental studies on hypersonic flows in the framework of transitional and near free molecular regimes. This free flow and continuous operation facility can provide a flow rate between 0 and 6 g/s depending on the power required from the pumping unit. The wind tunnel is composed of a first chamber acting as a plenum and a second serving as the test section. Under supersonic conditions, the plenum (length 2.25 m, diameter 1.2 m) is continuously fed by compressed gas from a reservoir. In the hypersonic regime, it is equipped with a nozzle supplied with nitrogen from a different reservoir. Figure 6.29 shows a schematic representation of the MARPHy wind tunnel in hypersonic configuration.

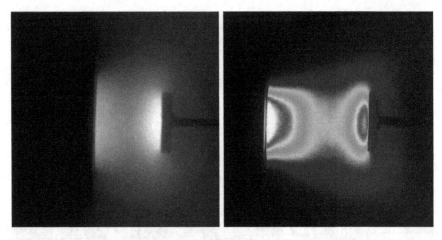

Fig. 6.26 Study of the interaction between the disk stagnation point and a supersonic plasma jet in the PHEDRA wind tunnel of the ICARE institute (© Institut ICARE)

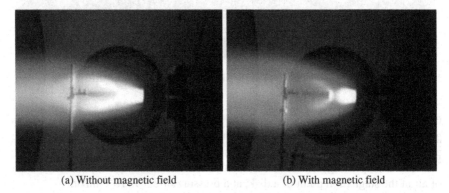

(a) Without magnetic field   (b) With magnetic field

Fig. 6.27 Influence of a magnetic field on the interaction between a truncated cylinder and a supersonic argon plasma jet. PHEDRA wind tunnel of the ICARE institute (© Institut ICARE)

Under supersonic conditions, it can be operated at Mach numbers of 2 and 4 with test section static pressures between 2 and 8 Pa. In a hypersonic regime, the Mach number can be set between 15 and 20 while the gas supplying the nozzle being heated to about 1300 K before expansion. Given the very low levels of static pressure in the test section, the use of a diffuser is essential.

The gas is heated up through a graphite electrical resistor before reaching the plenum and the throat of the nozzle which consists of a double-wall conical divergent and an interchangeable throat, the two elements being cooled by water. The diameter of the throat can vary between 1 and 3 mm depending on the desired Mach number and mass flow. The diffuser consists of a conical inlet followed by a cylindrical extension of 0.36 m in diameter and 2.2 m long downstream. Figure 6.30 shows a view of the test chamber of the MARPHy wind tunnel.

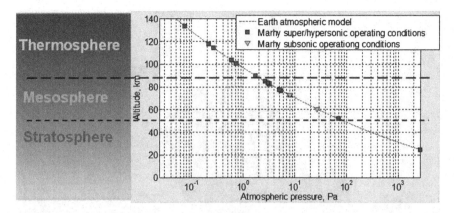

Fig. 6.28 Operational envelope of the MARPHy facility of the ICARE institute (© Institut ICARE)

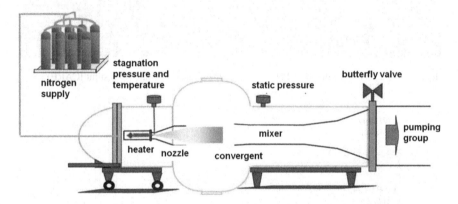

Fig. 6.29 Layout of the MARPHy wind tunnel in hypersonic configuration (© Institut ICARE)

The MARPHy wind tunnel is well suited for fundamental research on compressible rarefied flows, in particular on the aerodynamic and aerothermal behaviours of space probes and spacecraft.

6.3.6 Other Plasma Type Facilities

Airbus Safran Launchers (ASL) at the Issac site in France also has a set of facilities to characterise the behaviour of vehicles and their thermal protection during atmospheric re-entry missions. The different facilities cover a wide range of aerodynamic and aerothermal applications as shown in Fig. 6.31.

A set of plasma generators can produce subsonic, supersonic, or hypersonic flows representative of distinct re-entry phases. The generators use three types of plasma

Fig. 6.30 Test section of the MARPHy facility of the ICARE institute (© Institut ICARE)

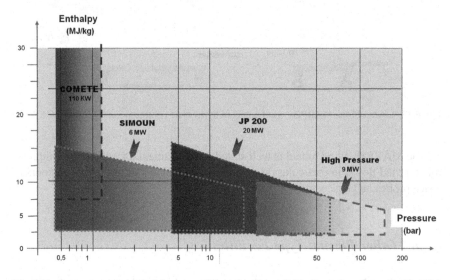

Fig. 6.31 Pressure ranges/enthalpies accessible on the Airbus Safran Launchers Issac site (© ASL)

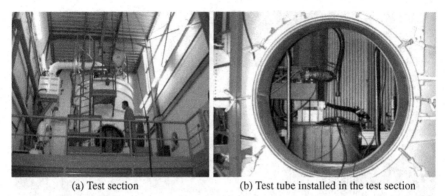

| (a) Test section | (b) Test tube installed in the test section |

Fig. 6.32 Test facility COMETE of Airbus Safran Launchers (© ASL)

torches to heat air, an air-argon mixture, or another gas such as N_2–CO_2 mixture that simulates the conditions of re-entry to Mars.

The **COMETE test facility** is a low power facility that uses an inductively generated plasma (electromagnetic waves are used to ionise the air) while the plasma torch being placed in an instrumented vacuum chamber (see Fig. 6.32a).

Models facing the generator outlet are subjected to hot flow over long periods of time (see Fig. 6.32b). The main features of the COMETE facility are:

– electrical power of 75 kW,
– stagnation enthalpy greater than 8 MJ/Kg (air),
– test duration ranging from a few seconds to 30 min,
– air, argon or nitrogen as operating gas,
– diameter of the subsonic jet of 80 mm,
– diameter of the tested articles of 50 mm.

The stresses subjected by the test articles vary from 30 to 400 mbar in pressure and from 300 to 3000 kW/m^2 in heat flux (cold wall conditions) under subsonic conditions.

The second technology is based on Huels-type arc heater and operates according to the schematic shown in Fig. 6.33. An electric current is sent into the upstream electrode and is transferred through the gas to a hollow downstream electrode through which the fluid flows. An external magnetic field is applied to rotate the arc foot and thus limit the erosion of the electrodes, the walls being cooled by a water circuit.

The **SIMOUN**, another test facility of ASL shown in Fig. 6.34, operates on a plasma torch principle generated by an electrically driven pneumatic and hydraulic system.

At the outlet of the plasma generator, a nozzle accelerates the plasma in order to generate representative conditions for model testing. A vacuum chamber allows high flow rates while the gases are being exhausted through a diffuser downstream of the test chamber. The square test chamber is instrumented with various optical means

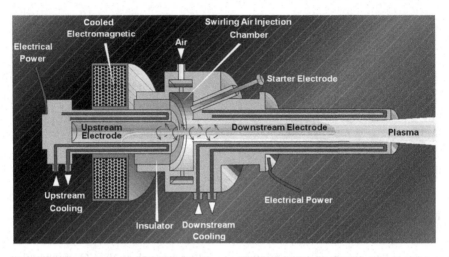

Fig. 6.33 Schematic diagram of Huels plasma generator technology (© ASL)

Fig. 6.34 SIMOUN testing facility of Airbus Safran Launchers (© ASL)

for flow diagnostics and study of ablation of thermal protection using normal high speed or thermal imaging infra-red cameras, pyrometer and etc.

The main features of the plasma generator are the following:

– electric power of 6 MW,
– total pressure ranging from 1 to 18 bar,
– total enthalpy ranging from 4 to 14 MJ/kg (air),
– test duration of a few seconds to 30 min.

Two types of tests can be conducted:

– an axisymmetric nozzle can accelerate the flow to Mach 4.5 on models of 50 mm diameter mounted on a mobile sting for the study of heating at the stagnation point,
– a semi-elliptical nozzle allows for Mach 5 flow to be established on flat plate configuration of 300 mm by 300 mm which could be also set at an angle of incidence.

The **JP200** test facility combines four Huels-type heaters to achieve even higher levels of perturbation on the models (see Fig. 6.35). The four plasma generators are coupled and discharged in a plenum allowing mixing and settling of the four streams. This chamber is connected to a nozzle that accelerates the flow. The nozzle may be axisymmetric to impose a stagnation point on models, or can be also adapted

Fig. 6.35 JP200 plasma test facility of Airbus Safran Launchers (© ASL)

to supply a square or rectangular test section. In the latter configuration, the effect
of a plasma flow on different types of surface coating materials can be studied by
placing them on the flat wall of the test section. The effect of pressure gradients on
these materials can be also tested on an inclined plate placed at the outlet of the test
section.

The main features of this facility are:

- electrical power of 4 times 5 MW,
- stagnation pressure ranging from 5 to 60 bar,
- dimensionless stagnation enthalpy ranging from 60 to 170 (air),
- test time is approximately 1 min, with the possibility of variation of the operating
  point during operation.

Since the plasma jet exhausts to the atmosphere the expansion ratios are limited
and the facility is used for:

- stagnation point test, where the test chamber is equipped with a contoured axisymmetric nozzle providing a uniform flow of Mach numbers of 1.7, 2, 4 or 2.6; the
  models subjected to the plasma jet are typically 50 mm in diameter and can experience a stagnation pressure of 5–50 bar, where pressure, heat flow, temperature
  and ablation measurements are performed;
- more global assessment in the closed square test section or inclined plate configuration where different nozzles with rectangular or square outlet sections allows
  for a Mach 1.7 or 2.4; the test samples, typically 200 mm long and 40 mm wide,
  can be subjected to pressures of 1–15 bar and heat fluxes of 1–25 MW/m^2; pressure, heat transfer, temperature (wall or material) or ablation measurements can
  be performed.

The **JPHP** facility shown in Fig. 6.36 allows for higher pressure testing (greater
than 100 bar), so as to ensure representativeness for most possible missions for the
vehicles concerned.

The main characteristics of this facility which also operates on Huels principle
are:

- electric power 9 MW,
- stagnation pressure greater than 100 bar,
- dimensionless stagnation enthalpy from 60 to 140,
- test duration of the order of one minute.

Axisymmetric nozzles can provide Mach 1.7, 2.4 or 2.6 flows. A multi-test rotary
arm permits successive positioning of up to eight samples or pressure or temperature
probes in front of the plasma jet (stagnation point configuration). Test specimens
typically have a diameter of 50 mm and their temperature and ablation rate can be
measured throughout the run by optical means (cameras) or more intrusive techniques
(thermocouples).

In Huels type generators, the length of the electric arc is free and can therefore
fluctuate, which limits the accessible powers. In the latest technology, the segmented
generator uses a predefined arc length to reach higher power levels. The schematic

diagram of such a system is given in Fig. 6.37. An arc, generated between two cooled electrodes, is stabilised in a channel made of many segments, electrically isolated from each other and cooled by a water circuit. The gases injected between the segments supply the test chamber while protecting them with the addition of colder air near the walls subjected to the thermal effects. The operation of the segmented generators which are more stable than the Huels type generators makes it possible to operate at higher powers.

Figure 6.38 shows a test facility that has implemented this technology. A seg-

Fig. 6.36 JPHP plasma test facility of Airbus Safran Launchers (© ASL)

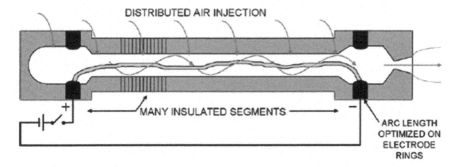

Fig. 6.37 Schematic diagram of segmented plasma generator technology (© ASL)

Fig. 6.38 Segmented plasma generator in operation (© Tekna)

mented plasma generator is used by the SCIROCCO wind tunnel at CIRA (see Sect. 6.3.5).

Chapter 7
Flow Visualisation Techniques

7.1 Earlier Contribution to Visualisation Techniques

In aerodynamics and more generally fluid mechanics one has the invaluable advantage of being interested in the physical phenomena that can be visualised directly, unlike other disciplines where only the consequences of the phenomena can be observed. While studying fluid flows, engineers and researchers have developed more or less sophisticated visualisation techniques to help them in their approach. Among the forerunners of aerodynamics, it is worth mentioning Étienne-Jules Marey (1830–1904) whose visualisations served as a starting point for the first modern wind tunnel, designed by Gustave Eiffel and installed at the foot of the tower in 1909. The device developed by Marey consisted of producing very thin and parallel smoke streams flowing vertically and sucked downwards by a fan (see Fig. 7.1). Then by placing an obstacle in the flow, the deviation in the path of the smoke while passing around the obstacle can be studied. Marey developed several devices, the latest version showing the speed of the flow of the smoke by taking a series of snapshots which is in fact a predecessor of the PIV. Smoke visualisations are still used to detect flow separations or vortices generated by the vehicle or models (see below).

Another very old visualisation method is to use tufts (silk, woollen or any other fibre) whose ends are glued to the surface of the model. Like a boat sail, the free end is aligned or flapping in the flow, thus indicating the direction of the flow and its fluctuations respectively, confirming the existence of unsteadiness. This technique is very economical and easy to implement and is still used to detect regions of separated flow (see Fig. 7.2).

Optical techniques are of primary interest for the visualisation of shock phenomena in high-speed flows showing zones of rapid expansion and compressions, boundary layers and wakes. There are many visualisation methods that have emerged in aerodynamics, but will be not all treated here. In the following, we limit ourselves to the visualisation techniques mostly used in today's applications.

© Springer Nature Switzerland AG 2020   165
B. Chanetz et al., *Experimental Aerodynamics*,
Springer Tracts in Mechanical Engineering,
https://doi.org/10.1007/978-3-030-35562-3_7

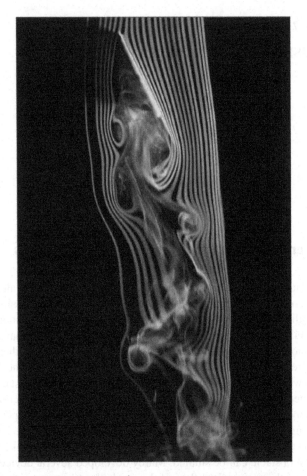

Fig. 7.1 Smoke visualisation realised by Étienne-Jules Marey (© CFCA)

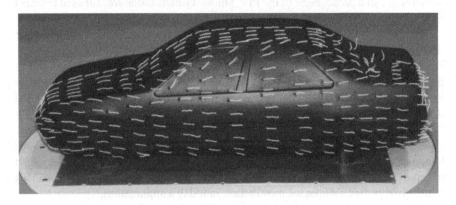

Fig. 7.2 Tufts visualisation of separated flow region on a car (© PSA Peugeot-Citroën)

7.2   Surface Flow Visualisations

The purpose of the surface flow visualisations is to highlight limit streamlines or wall streamlines that are defined as the limit of a streamline when the distance y to the wall tends to zero (see Fig. 7.3). For a Newtonian fluid, it is shown that at the limit $y \to 0$ the direction of the velocity vector becomes collinear with the skin friction vector at the wall, hence the notion of skin friction line defined as a line tangent to the skin friction vector. Wall streamline (or limiting streamline) and skin friction line are in principle identical concepts. However, from a physical point of view, it is preferable to adopt the concept of a skin friction line that is unambiguously defined from a measurable physical quantity, rather than the concept of a limiting streamline arising from theoretical boundary conditions, sometimes questionable. The set of skin friction lines covering an obstacle is called the surface flow pattern.

7.2.1   Surface Oil Flow Visualisation

To visualise the surface flow pattern, the model is coated with a viscous liquid before blowing the air over it. The assumption of these techniques is that the traces or lines of striations, resulting from the shearing action exerted by the flow on the viscous product can be identified with skin friction lines. This assumes that the coating is runny (thin) enough not to influence the flow in the near-wall region while being sufficiently viscous (thick) to not be flushed out. Visualisation products range from white spirit to glycerine and silicon oils which are often intimately mixed with white or fluorescent pigments such as Titanium dioxide and Day-Glo powder. Different colours can thus be used to visualise a more complex flow pattern as shown in Fig. 7. 4.

Fig. 7.3 Velocity profile in a three-dimensional boundary layer

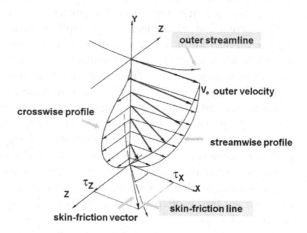

Fig. 7.4 Visualisation by colourful coatings of the rear part of a business jet model (© ONERA)

There is no general "recipe" for a good coating. The quality of the visualisation depends on the nature and the viscosity of the oil, the dye, the technique of recording the images, as well as the velocity, pressure and temperature of the flow to be studied, and the skill of the experimenter. The mode of operation of the wind tunnel (continuous or intermittent) can also affect the results.

In two dimensional, planar or axisymmetric, flows, surface visualisations are useful for verifying the two-dimensional character of the flow and for detecting any separated regions. For three-dimensional configurations, surface flow visualisations are often indispensable, the surface flow pattern being the imprint of the organisation of the external flow, often much more complex than in two-dimensional configurations (see Fig. 7.5). In most cases of practical interest, the surface flow pattern of a three-dimensional field has the peculiarities of revealing the presence of critical points and separation lines. The rational interpretation of such pattern is through critical point theory which leads to distinguish critical points such as nodes, foci or saddle points, and dividing streamlines which are skin friction lines passing through a saddle point. The dividing streamlines may be due to separation or attachment depending on the behaviour of the flow in the vicinity. It is the careful inspection of the surface flow pattern, with the detection and identification of its critical points and lines of separation or attachment, which gives a first idea of the formation of a region of separated flows rolling-up into vortex structures. Thus, a good visualisation of the surface flow is often a prerequisite for further exploration of the flow field by probing techniques.

An example of a complex surface flow pattern presented in Fig. 7.6a shows the visualisation of the skin friction lines on part of the central body of an aerospike

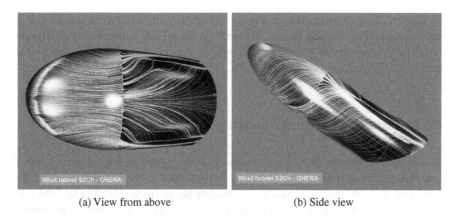

(a) View from above   (b) Side view

Fig. 7.5 Surface flow pattern on a flattened ellipsoid body (© ONERA)

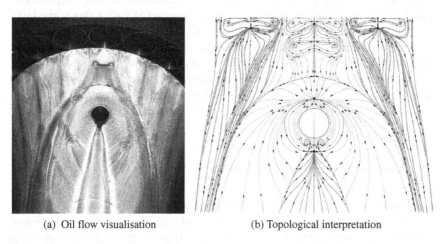

(a) Oil flow visualisation   (b) Topological interpretation

Fig. 7.6 Surface flow pattern on the central body of an aerospike-type nozzle (© ONERA)

type launcher nozzle. It shows the structures resulting from the impact on the central body of the jets emerging from the propulsive nozzles and the separation caused by the emergence of a transverse jet used for thrust vectoring. Figure 7.6b gives a topological interpretation of the near wall flow field. For locally flat surfaces, the validation of the critical points suggested by visual analysis can be carried out on the basis of two-dimensional analytical models capable of representing the velocity field induced by critical points such as nodes and foci (of attachment and/or separation type). The models are based on the mathematical singularities of source and vortex types that satisfy Laplace's equation for the velocity potential. The analogy is purely qualitative and does not imply any hypothesis about the origin or the nature of the flow. The comparison of the number and the position of the saddle points detected experimentally with the number and the position of the saddle points resulting from the numerical treatment makes it possible to identify the nature (of attachment or separation) of the singular points and singular lines. The approach, coupled with an interactive numerical model, can be used as a quick way to validate the visual analysis of the surface flow visualisations.

7.2.2 Visualisation by Sublimating Product

This method, consist in depositing a product on the model which sublimates under heat transfer from the flow; it is mainly used to detect the laminar to turbulent transition. Indeed, the heat transfer coefficient being higher in turbulent than in laminar flows, sublimation of the product will be faster on the parts of the model where the boundary layer is turbulent (this temperature difference is also used by infrared techniques and TSP, see Sect. 9.5). The most common product is naphthalene diluted with acetone. Before testing, the model is covered with a thin white layer of naphthalene which gradually disappears with the establishment of the flow, the turbulent parts losing their whiteness before the remaining laminar regions. Since the entire product will sublimate after a while, the snap shots should be taken at the right instants. Infrared thermography, presented in Sect. 9.5.3 as a quantitative method for measuring surface heat flux, is also used to visualise the laminar to turbulent transition (see Sect. 1.7).

7.3 Visualisation in Water Tunnels

Water tunnels are valuable tools for visualising low velocity flows weakly dependent on the Reynolds number (see Sect. 3.4). The technique involves the introduction of dye tracers into the flow in the form of filaments of a liquid whose density is very close to that of water, such filaments may be of different colours to distinguish the various structures of the flow. The injection is done either upstream of the visualisation region or through holes located on the model. Fluorescein is sometimes used as tracer and gives spectacular images. Figure 7.7 shows a visualisation by coloured dye filaments of the flow on a model of combat aircraft highlighting the breakdown of the intense vortices forming on the upper surface of the delta wing.

Another technique is to generate small diameter air bubbles by means of a foaming product upstream of the model (see Fig. 7.8), these bubbles are illuminated by a light sheet to visualise a plane of the flow. A snapshot is then taken with a certain exposure time and trajectories of the bubbles are captured as trails.

The images obtained are very instructive. The method is the origin of PIV which experienced a major development thanks to the progress in the fields of laser-optics and signal processing. Figure 7.9 shows a view, from downstream, of the flow in a vertical plane highlighting the vortices on the upper surface of the wings of a Concorde model.

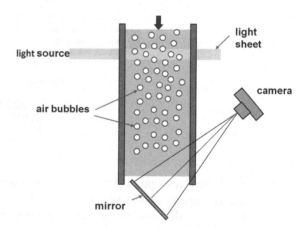

Fig. 7.7 Vortex breakdown over the delta wing of a combat aircraft (© ONERA)

Fig. 7.8 Visualisation by the method of air bubbles in a water tunnel

7.4 Laser Tomoscopy or Laser Sheet Visualisation

The principle of this method is to observe the light scattered by particles injected into the flow which is illuminated by a sufficiently strong laser. Water droplets or smoke particles are produced by atomisation of appropriate oil. In most applications, the light sheet is produced by a laser beam passing through a glass rod (cylindrical lens) or a combination of lenses (see Fig. 7.10). Very meaningful images can be obtained showing vortices, boundary layers or shock waves.

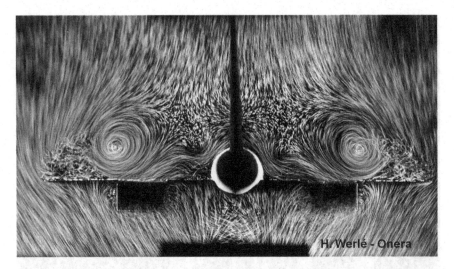

Fig. 7.9 Air bubble visualisation past a Concorde model, seen from downstream (© ONERA)

Fig. 7.10 Principle of the laser sheet visualisation

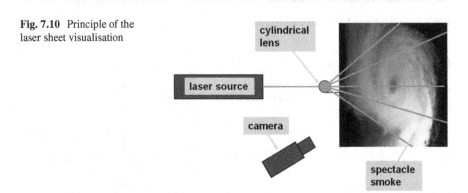

A more complete description of the field can be made by moving the laser plane or using multiple laser sheets as shown in Fig. 7.11, which highlights the vortices forming above a missile body at incidence, in a supersonic flow.

7.5 Visualisation by Optical Imaging

7.5.1 Schlieren and Shadowgraph Techniques

The Schlieren and shadowgraph techniques are used to visualise flows with sufficient variations in density, ρ, to cause a significant modification of the refractive index of the gas. These methods are thus recommended for high-speed flows, from high

Fig. 7.11 Laser sheet visualisation of the vortices above a missile body at incidence at Mach 2 (© ONERA)

subsonic to hypersonic regimes. In two-dimensional flows (planar or axisymmetric), these techniques allow an identification of the structure of the flow obtained from compression and expansion waves, shock waves and high shear regions (such as boundary layers and mixing layers). The Schlieren technique can be applied in very low velocity water flows by inducing a variation of the refractive index by localised surface heating.

The shadowgraph technique: this technique consists of placing a point light source and a photo-diode (light sensor) on either side of a disturbed flow (see Fig. 7.12).

The method responds to the second spatial derivative of the refractive index of light which is proportional to the density, and therefore to the temperature and/or pressure.

Figure 7.13 shows a short exposure time shadowgraph of the model of a space probe in flight in the ISL ballistic tunnel highlighting the bow shock wave and turbulent structures in the wake, as well as the acoustic waves generated by these structures and their propagation in the neighbouring supersonic flow.

Figure 7.14 shows a coloured shadowgraph of the flow past the nose of a model equipped with an aerospike aiming at reducing the drag at transonic and supersonic speeds.

Particular optical devices, such as conical shadowgraph, have been developed for observation in planes perpendicular to a direction along which the flow exhibits conical similarity, as in a shock wave/boundary layer interaction induced by a swept obstacle.

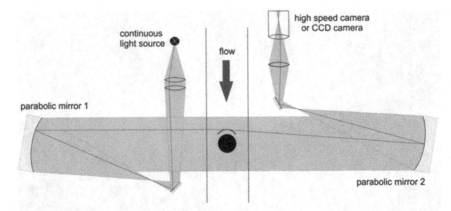

Fig. 7.12 Principle of a shadowgraph optical bench

Fig. 7.13 Shadowgraph of a space probe model fired in the ISL aeroballistic tunnel (© ISL)

Schlieren technique: this technique is slightly different; the light which is not deflected by the gradient of the refractive index of the flow is suppressed by the use of a knife (see Fig. 7.15). The method highlights the first derivative of the refractive index of light, directly related to the density gradient. The Schlieren technique is more sensitive than the shadowgraph because it offers greater dynamics of the grey

Fig. 7.14 Coloured shadowgraph performed in the S3MA wind tunnel of the ONERA Modane-Avrieux centre (© ONERA)

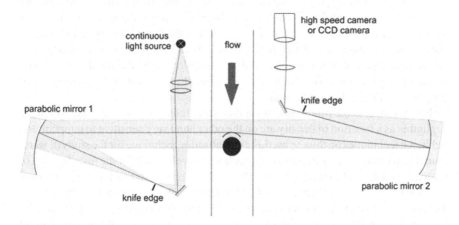

Fig. 7.15 Principle of a Schlieren setup

scale of the image. The vertical or horizontal orientation of the knife edge makes it possible to resolve the direction of the density variation.

Figure 7.16 shows a short exposure time Schlieren picture of the flow produced by an aerospike type nozzle placed in a supersonic flow at Mach 2 in the S8Ch wind tunnel of the ONERA, Meudon.

Fig. 7.16 Schlieren picture of the flow produced by an aerospike nozzle (© ONERA)

In conventional Schlieren optical benches, the domain of integration is the region between the two mirrors installed on each side of the test section, which also collimates the light beam passing through the flow. The system gives a clear picture of the properties of the flow passing the beam, the images being difficult to interpret in a three-dimensional flow. In focussed Schlieren systems, the object is out of the focusing area after a short distance. Thus, effects of the density variation in a narrow slice of the flow normal to the optical axis are visualised. If the flow is steady, or if the average characteristics are observed in a turbulent flow, the focal plane can be moved through the test volume to obtain a three-dimensional description made of images recorded at different instants.

For incompressible fluids such as liquids, variations in refractive index which are negligible as a function of density are on the other hand very sensitive to temperature. Schlieren technique can thus be used as a visualisation technique for the study of water tunnel flows thanks to a surface heating technique, well suited for the investigation of unsteady and three-dimensional vortex flows. The heat transfer technique (or thermal marking) consists of equipping a wall with a copper film mounted in a flush insert. This metal element is in contact with an electrical resistor inside the model heated by Joule effect, a power supply unit allows adjusting the intensity of the electric current through the resistor. Figure 7.17 shows a visualisation by surface heating of the apex vortices of a tapered body.

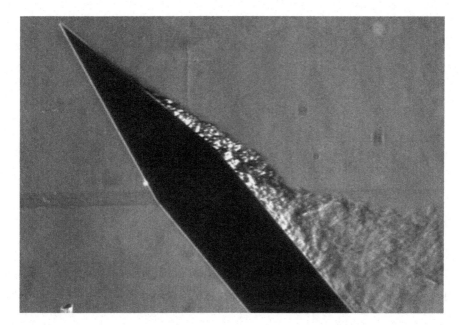

Fig. 7.17   Schlieren by thermal marking technique on a cone in a hydrodynamic tunnel (© ONERA)

7.5.2   *Interferometry*

Also based on refractive index variations in compressible flows, interferometry is a powerful means for quantitatively analysing transonic and supersonic flows. Its principle and implementation are presented in detail in Sect. 11.2. Interferometry can also be used as a means of visualising compressible flows (see Fig. 7.18). However, its more delicate use makes the Schlieren and shadowgraph techniques more preferred for simple visualisations.

7.5.3   *Differential Interferometry*

The differential interferometry developed at ISL in the eighties is more sensitive to the density gradient than shadowgraph and Schlieren techniques. Instead of being totally separated, as in conventional differential interferometers, the two interfering light beams pass through the phase object with only a small lateral shift relative to each other, the beam being separated by a double Wollaston prism consisting of a birefringent crystal such as quartz or calcite.

Figure 7.19 presents the principle of a differential interferometry bench, two perpendicularly polarised light beams are expanded and separated by a few tenths of a millimetre by the Wollaston prism before passing the flow field as a collimated

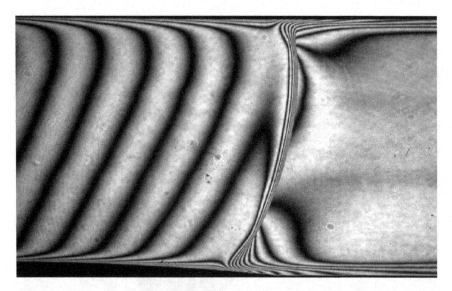

Fig. 7.18 Interferogram of a transonic flow with shock wave (© ONERA)

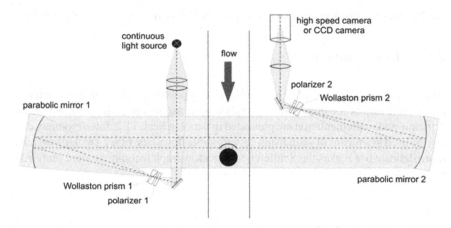

Fig. 7.19 Principle of a differential interferometry bench

light beam. They are then gathered by a second Wollaston prism and then analysed by a polariser and focused on the CCD camera by a lens system. Variations in the flow density generate different optical path lengths between the two split light beams, giving an interference pattern on the CCD camera. The bench can be adjusted to obtain a pattern of fringes or a single fringe giving a uniform light intensity distribution. In this case, the images look like Schlieren images, the density gradient field being visualised in terms of the distribution of light intensity on the interferograms.

Such an interferometer has significant advantages over other visualisation techniques:

- it is relatively insensitive to vibrations and imperfections of the optics
- it offers a great flexibility in the visualisation of different phenomena, in particular in very low-density gases, the sensitivity being adjustable by the choice of the Wollaston prism angle.

Figure 7.20 shows a false-colour image, obtained by differential interferometry of the separation of two half-cylinders surrounding a sting which produced a shock in the tunnel at Mach 4.5, for altitude conditions of 13.9 km.

To highlight the differences in sensitivity between the methods, Fig. 7.21 compares the three visualisation techniques used in the ISL shock tunnel. The case presented corresponds to the flow past a sphere for the Mach number of 10 at altitude conditions of 50 km.

Fig. 7.20 Image obtained by differential interferometry of the separation of two half-cylinders (© ISL)

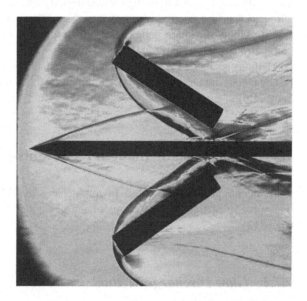

(a) Shadowgraph   (b) Schlieren   (c) Differential interferometry

Fig. 7.21 Visualisations of the flow past a sphere at Mach 10 for an altitude of 50 km (© ISL)

7.6 Short Exposure Time Visualisation

The transient and unsteady nature of many aerodynamic phenomena (inlet starting of a supersonic nozzle, flying projectile, explosion, turbulence, propagation of waves, etc.) has led to the development of visualisation methods in which the phenomenon is frozen. From the beginning of their development, shadowgraph, Schlieren technique and differential interferometry were used in the case of high-speed flows with the main difference being the technique of capturing the image on a sensor. In those days, the tests were performed in total darkness using spark system to generate a bright spark with a very short exposure time (a few ns), which freezes the flow on a photographic plate with almost no blurring. Then, spark systems were used in series and a high-speed drum camera with the photographic film attached to the drum which allowed capturing several images during a test. Typically, in hypersonic flow, 8 images were recorded for 3 ms of flow. This recording technique was abandoned in the 2000s with the advent of high-speed digital cameras. The most efficient models allow an aperture opening time of less than 300 ns, which is comparable to the exposure time of a spark gap. The acquisition rate can be up to 25,000 frames per second for a resolution of 1280 by 800 pixels.

Methods such as PIV, based on the particle displacement, also require an almost instantaneous image obtained by illuminating the field with a pulsed laser (see Sect. 11.6).

7.7 Visualisation by Induced Light Emission

These techniques, mainly used in low-density hypersonic flows, are based on the emission of light by the atoms of the gas under the effect of ionisation (see Chap. 12).

7.7.1 Glow Discharge Method

The gas is ionised by an intense electric field, which produces a plasma under the effect of the excitation of the nitrogen molecules. This results in a light emission whose intensity is a function of the local density. The process, which only works in very low density flows, highlights shock waves, boundary layers, mixing zones,

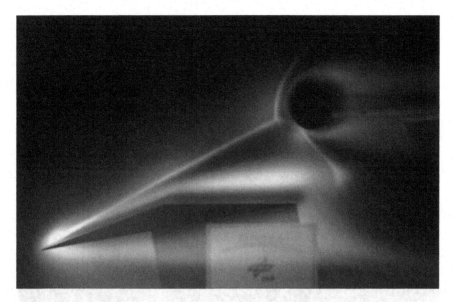

Fig. 7.22 Glow discharge visualisation of a shock-shock interference at Mach 10 (© DLR)

etc. Figure 7.22 shows a visualisation of the field produced by the intersection of an oblique shock wave generated by a wedge and the shock forming in front of a cylinder (experiments performed by DLR in the ONERA R5Ch wind tunnel).

7.7.2 Electron Beam-Induced Fluorescence (EBF)

This process, which uses light emitted by nitrogen molecules while de-energising after intense electron beam bombardment (30 kV–0.1 mA), is also used for quanti-tative measurements (see Sect. 12.7). Figure 7.23 shows an EBF visualisation of a Mach 10 flow past a Martian probe. The image of the field is obtained by scanning the electron beam at a frequency of 50 Hz.

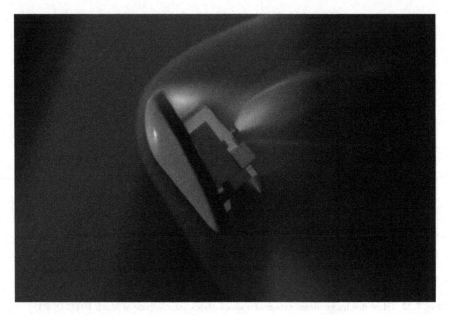

Fig. 7.23 Electron beam fluorescence visualisation of a Mach 10 flow past a Martian probe (© ONERA)

Chapter 8
Measurement of Aerodynamic Forces and Moments

8.1 The Aerodynamic Forces and Moments

The presence of the flow around a vehicle exerts aerodynamic forces and moments, which act at the centre of pressure. The problem posed to the aerodynamicist is to determine these forces and their point of action relative to the centre of gravity, G, of the aircraft. The equilibrium of the aerodynamic forces is thus reduced to a force and a moment applied at G (see Fig. 8.1). These are the necessary information required to calculate the aircraft's trajectory and its motion around the centre of gravity G.

In the wind tunnel, the loads on a model can be measured by multi-components balances, with respect to a reference position, the centre of reduction of the forces whose position must be provided to transport the moment to any other point (frequently the mean aerodynamics centre which is at around the quarter chord position in the case of standard aerofoil). This allows the calculation of the aerodynamic forces and moment coefficients which can be therefore extrapolated to the real vehicle.

The aerodynamic coefficients can be decomposed into the Euclidean space, resulting into three forces and three moments as shown in see Fig. 8.2. The decomposition being done according to various coordinate systems:

- the aerodynamic axis, linked to direction of the headwind,
- the aircraft axis, linked to the aircraft or the model,
- the wind tunnel axis, linked to the wind tunnel in which the tests are carried out,
- the balance axis, linked to the system used to measure the aerodynamic forces and moments.

Here the aerodynamic coordinate system, whose axes are thus defined, will be considered:

- the longitudinal axis $\overrightarrow{X}_a$ aligned with the velocity vector $\overrightarrow{V}_\infty$, positive forward,
- the vertical axis $\overrightarrow{Z}_a$ normal to the horizontal plane $\overrightarrow{X}_a$, positive upward,
- the transverse axis $\overrightarrow{Y}_a$ normal to $\overrightarrow{X}_a$ and $\overrightarrow{Z}_a$, positive towards the right hand side of the plane (starboard).

© Springer Nature Switzerland AG 2020
B. Chanetz et al., *Experimental Aerodynamics*,
Springer Tracts in Mechanical Engineering,
https://doi.org/10.1007/978-3-030-35562-3_8

Fig. 8.1 The aerodynamic forces acting at the centre of gravity of an aircraft

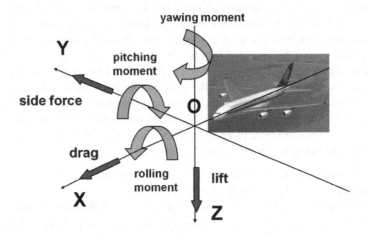

Fig. 8.2 Forces and moments constituting the aerodynamic coefficients

In this system, the force $\vec{F}$ is decomposed into:

- a component along $\vec{X}_a \rightarrow$ drag F_x,
- a component along $\vec{Z}_a \rightarrow$ lift F_z,
- a component along $\vec{Y}_a \rightarrow$ side force F_y.

In a similar way, the moment $\vec{M}$ is decomposed into:

- a component along $\vec{X}_a \to$ rolling moment M_x,
- a component along $\vec{Z}_a \to$ yawing moment M_z,
- a component according to $\vec{Y}_a \to$ pitching moment M_y.

If the aerodynamic force components are known in a certain system of axes T_1 its components in any other system T_2 are deduced by the transformation matrix $[R]$ of the kind:

$$\begin{bmatrix} X' \\ Y' \\ Z' \end{bmatrix}_{T_2} = [R] \begin{bmatrix} X \\ Y \\ Z \end{bmatrix}_{T_1}$$

A major challenge for aerodynamicist is to identify the positions at which the aerodynamics forces are acting. This has serious implication for flight mechanics as the stability and control of the aircraft relies on this position, relative to the centre of gravity of the overall aircraft. This introduces the concept of the neutral point, which is the position of the mean aerodynamic centre at which the aircraft is stable when responding to a sudden change in attitude (incidence for example). During the design, a static margin, which is the distance between the centre of gravity and the neutral point, is defined. For an aircraft to be stable the centre of gravity should always be ahead of the neutral point and hence a positive static margin, expressed as a percentage of mean aerodynamic chord.

The measurement of the forces exerted on a model is carried out by means of aerodynamics balances comprising of a set of strain gauges which measures a deflection due to the forces acting on the vehicle, and using the simple Hooke's law the forces can be determined. These balances have very varied architectures depending on the vehicle being tested; the smallest being only a few cubic millimetres in volume and can be installed in the model itself.

8.2 Aerodynamic Balances

8.2.1 Forces and Strain Gauges

The six components of the aerodynamic coefficients are determined by means of force balances whose principle is based on the measurement of the deflection of elements in the shape of the plates or beams. These deformations are measured by load cells equipped with extensometers or strain gauges, which sense the deflection of a material and transmit it as a variation in electrical resistance or voltage (the resistance being directly proportional to the deflection,). The strain gauge consists of an insulating flexible backing, supporting a metallic foil pattern (a few μm thick) similar to a printed circuit (by lithography or etching) as shown in Fig. 8.3. While

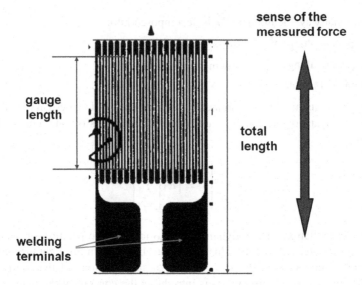

Fig. 8.3 Typical resistive foil strain gauge

measuring small deformations the gauges operate in the elastic regime under Hooke's law.

Under the action of a force F, the length L of a beam is extended by ΔL and this is defined as the strain of the material expressed as:

$$\epsilon = \Delta L / L$$

The stress experienced by the beam can be expressed as:

$$\sigma = F / A$$

where A is the cross sectional area of the beam. The stress and strain can be related by:

$$\sigma = E \times \epsilon$$

where E denotes the Young's modulus of the material which is fixed at constant temperature. The measurement of the strain ε thus makes it possible to determine the stress σ hence the force F.

In a resistive type strain gauge, where the resistive element is either a wire or a semiconductor, the elongation resulting from the application of a force causes a variation of electrical resistance:

$$\frac{\Delta R}{R} = k \frac{\Delta L}{L}$$

and the so called gauge factor is the ratio:

$$k = \frac{\Delta R/R}{\Delta L/L} = \frac{\Delta R/R}{\varepsilon}$$

The variation of resistance, due to the deflection is measured by inserting the gauge as one of the resistance branches in a Wheatstone bridge, but direct measurement is rarely done. In most applications, four gauges fixed to the deforming element of the balance constitute the Wheatstone bridge which operates under the forces or the moments exerted on the beam by the aerodynamic loads from the strut or model supports. The components of each force are then deduced from relations obtained by applying the Wheatstone bridge equation, which will not be given here. Thus using the arrangements shown in Fig. 8.4, it is possible to measure a force applied to an unknown point from the bending moment in a beam and therefore determine the aerodynamic forces and moments, including the rolling moment, using two bridges. For the direct drag measurement, the arrangement in Fig. 8.5a can be used. This rather poorly sensitive method is not suitable for measurement of low drag, where the method based on a deforming parallelogram is preferred. This arrangement further amplifies the forces to be measured and hence the accuracy of the measurement is increased (see Fig. 8.5b).

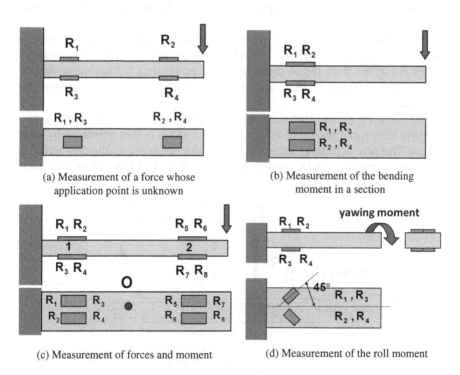

(a) Measurement of a force whose
application point is unknown

(b) Measurement of the bending
moment in a section

(c) Measurement of forces and moment

(d) Measurement of the roll moment

Fig. 8.4 Arrangements of strain gauges mounted in a Wheatstone bridge

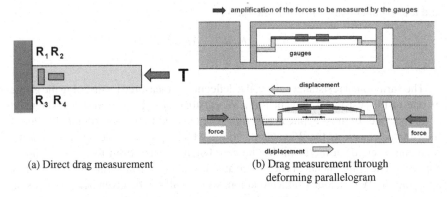

(a) Direct drag measurement   (b) Drag measurement through deforming parallelogram

Fig. 8.5 Strain gauges arrangement for the measurement of the drag

8.2.2 Sting Type Force Balance

The gauges are fixed to the deformable measurement element of the balance which is linked downstream to the sting supporting the model. The forces and the moments to which the model is subjected are transmitted to the rear end of the fixed sting and this causes the measurement piece to deform under the aerodynamic load (see Fig. 8.6).

The forces determined with respect to the balance virtual centre can then be transposed to any other point directly linked to the model. The schematic representations in Fig. 8.7 show the arrangement of the gauges on a balance measuring the six components of and Fig. 8.8 shows the ONERA 72C sting balance based on this principle.

The accuracy of such balances are affected by interactions or coupling to a response from a component of force or moment which is then registered by other gauges. This results from an imperfect decoupling: for example, the drag force, X, may cause a response on the bridges measuring Y, Z, L, M, N. These are due to

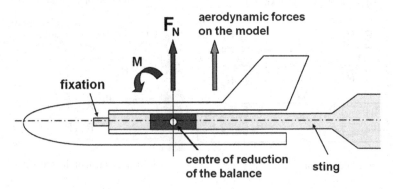

Fig. 8.6 Installation of a sting balance

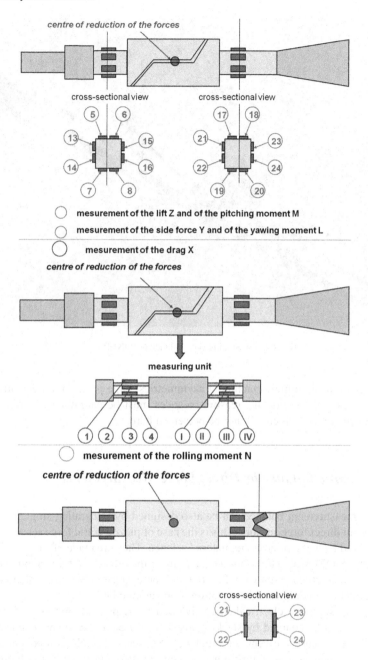

Fig. 8.7 Arrangement of the gauges on a six-component sting type balance

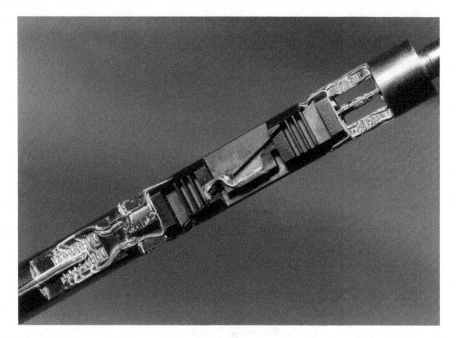

Fig. 8.8 ONERA 72C six-component sting type balance (© ONERA)

geometrical uncertainties, machining asymmetries, gage positioning, deviations or elastic deformations. For this reason, balances must be calibrated according to a complex procedure which will not be presented here.

8.2.3 Force Balance by Direct Force Measurements

A force measurement balance can be also designed by integrating single or multi-component direct force sensors. This is the case of piezoelectric force sensors which has evolved from the progress in material science. They are made of a material, such as quartz (SiO2), which develops a charge under the effect of a force applied on the faces of the crystal. As shown in Fig. 8.9, the force causes a longitudinal effect or a shear effect depending of the orientation of the atomic lattice.

The corresponding electrical charge being directly proportional to the force, the sensitivity can be adapted by stacking a series of sensitive elements connected in parallel, to increase the sensor's efficiency. Similarly, a multi-component sensor can be designed by integrating sensitive cells in different directions. Figure 8.10 illustrates the assembly of load cells for the measurement of normal and shear forces. Each force component is measured by a pair of piezoelectric cells, all connected in parallel to the whole assembly for the measurement of the 3 components in the X, Y and Z directions.

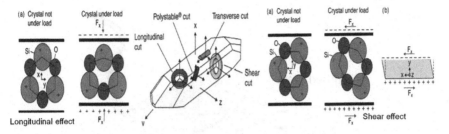

Fig. 8.9 Principle of longitudinal and transverse piezoelectric effects (© Kistler)

Fig. 8.10 Three components piezoelectric sensor (© Kistler)

The piezoelectric effect results in a linear relation which links the electric charge with the force applied, by the relation:

$$Q = S \times F$$

where F is the force in Newton (N), Q the charge in picocoulomb (pC) and S the sensitivity in pC/N (of the order of 2.3 pC/N in longitudinal load, 4.6 pC/N in transverse load).

The intrinsic electrical conductivity of a sensor is equivalent to the electrical charge under the effect of a load. Two principles are used: either an impedance conversion is made in situ in the sensor which then emits a voltage proportional to the force, or by transfer of the load of the sensor by means of a charge amplifier.

The force measurement balances thus consist of a fixed base and a load plate, often connected by a set of three components sensors (generally 3 or 4). The forces are set at equilibrium when the model is connected to the base plate and this counteracts the load due to the weight of the model. Figure 8.11 shows an assembly of a measurement rig where 4 three-component sensors are connected in series to form an assembly for the measurement of 3 force components along the X, Y and Z directions.

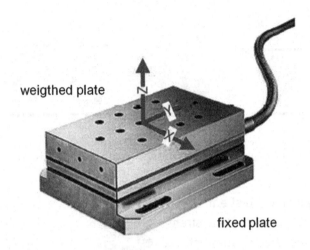

Fig. 8.11 Three-component load cells (© Kistler)

Figure 8.12 shows a wind tunnel setup for the study of the transient process of starting a supersonic nozzle. Using three 3-components sensors, whose outputs are measured individually, the complete aerodynamic load acting at the virtual centre of the balance is measured.

Therefore, the aerodynamic forces acting on the nozzle can be expressed as:

$$T = M * F$$

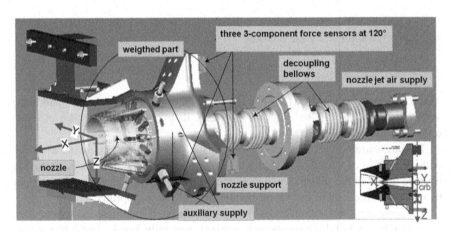

Fig. 8.12 Six-component balance system for the measurement of forces during the starting process of a supersonic nozzle. Tests in the R2Ch wind tunnel at ONERA, Meudon (© ONERA)

Fig. 8.13   Kistler 9327C force sensors and other models (© Kistler)

where F is the vector of the 9 force measurements and T the 6-component aerodynamic forces. The matrix M is calculated by a calibration of the balance using known forces along the 3 directions in space and inversion of the calibration matrix.

This type of sensor allows for rigid balances with minimal deformation, thus ensuring the precise attitudes of the wind tunnel model (incidence, side-slip and yaw). Combination of each component of the balance sensors with a load amplifier whose counter-reaction offers an adjustable time constant makes it possible to measure steady forces; which is referred as quasi-static measurements. This combination also allows for dynamic measurements whose bandwidth can reach a few thousands Hz. Examples of force sensors, two and three components, are given in Fig. 8.13.

8.2.4 Balance for Ground Vehicle

Measurement of the forces and moments exerted on an automobile is conducted by mounting the vehicle on the horizontal floor of the test section. The vehicle is connected by the wheels to a mobile frame and the translation and rotation motions are controlled and measured in an orthonormal frame relative to a fixed reference. The movable frame is connected to a fixed support through calibrated load cells, allowing direct measurement of the forces exerted on the vehicle. In the case of a six-component balance, the mobile and fixed frame connection is made using six load cells (see Fig. 8.14). A first load cell C_1 directed along x allows the measurement of the drag force. Two other load cells, C_2 positioned at the front and C_3 at the rear of the model in the direction y provide the front and the rear side forces, the total lateral force and then the yaw moment. The three remaining load cells C_4, C_5 and C_6 oriented vertically, are for the measurement of the overall lift force, F_z, the forward, $F_{z_{avt}}$ and rear $F_{z_{arr}}$ lift forces, the roll moment, M_x, (around the x-axis) and the pitch moment, M_y, (around the y-axis).

Let F_1, F_2, F_3, F_4, F_5 and F_6 denote the forces associated with the load cell C_1, C_2, C_3, C_4, C_5 and C_6 connected to the links B_1, B_2, B_3, B_4, B_5 and B_6 respectively

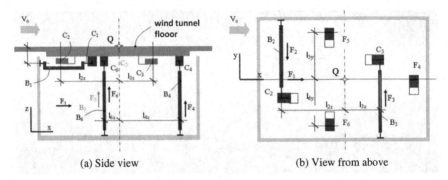

(a) Side view   (b) View from above

Fig. 8.14 Schematic diagram of a ground vehicle force balance

and as shown in Fig. 8.14. For this configuration, the drag, side force, lift and roll, pitch and yaw moments, can be thus determined. The force measured by load cell C_1 is directly linked to the drag force:

$$F_x = F_1$$

Load cells C_2 and C_3 transmits the forces F_2 and F_3, the side force and the yaw moment acting at point Q on the vehicle are given by,

$$F_y = F_3 - F_2$$

$$M_z = l_{3x} F_3 + l_{2x} F_2$$

The load cells C_4, C_5 and C_6 are then used to determine the lift force at the front and rear respectively and the total lift force from the relationships:

$$F_{z_{av}} = F_5 + F_6$$

$$F_{z_{ar}} = F_4$$

$$F_z = F_{z_{av}} + F_{z_{ar}}$$

The roll and pitch moments are given by:

$$M_x = l_{5y} F_5 - l_{6y} F_6$$

$$M_y = l_{6x}(F_5 + F_6) - l_{4x} F_4$$

Fig. 8.15 The force balance of the S2A wind tunnel (© Soufflerie S2A)

Figure 8.15 shows the force balance equipping the GIE S2A wind tunnel (see Sect. 3.3.3). The load cells are calibrated with known weights before each test.

8.3 Drag Determination from Wake Survey

The aerodynamic forces on a model can be determined indirectly by applying the fundamental equations of fluid mechanics: conservation of mass, momentum and energy. This method is mainly used to determine the drag, which is small compared to the lift of an aircraft and the magnitude of which is sometimes at the limit of balances accuracy. In addition, the method allows an analysis of the sources of drag, useful to optimise the design of an aircraft wing, for instance.

Consider the control volume containing the vehicle shown in Fig. 8.16 where (S_3) is a streamtube large enough for the pressure on its surface to be constant and equal to the pressure of the undisturbed upstream flow, (S_1) an upstream plane on which the flow is of uniform of pressure P_∞ and velocity $\overrightarrow{V}_\infty$, (S_2) a plane downstream, where the pressure P and velocity $\overrightarrow{V}$ are (non-uniform), and such that at large lateral distance P and $\overrightarrow{V}$ are becoming again equal to P_∞ and $\overrightarrow{V}_\infty$. The plane (S_2)

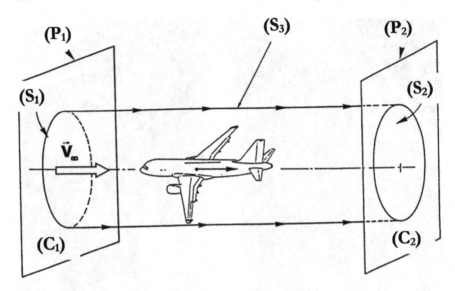

(P₁) **(S₃)** **(P₂)**

(S₁) **(S₂)**

$\vec{V}_\infty$

(C₁) **(C₂)**

Fig. 8.16 Control volume for drag evaluation from model wake survey

incorporates the vehicle wake in which the velocity and pressure vary. Therefore the
drag force, T, on an isolated vehicle placed in a flow of uniform upstream velocity
and pressure can be expressed as:

$$T = \frac{\vec{V}_\infty}{\left|\vec{V}_\infty\right|} \int_{S_2} \left[(P - P_\infty) + \left(\rho\vec{V} \cdot \vec{n}\right)\left(\vec{V} - \vec{V}_\infty\right)\right] dS$$

where S is the global domain containing the vehicle. This equation is known as
Oswatitsch's relation.

The terms of the integral being zero on the planes S1 and S3, the drag is deduced
from a survey of the flow in the downstream plane, S_2 (generally a plane nearly
normal to the upstream velocity vector). The velocity deficit can be translated into
stagnation pressure loss (or energy loss per unit volume in J/m^3) and the magnitude
of which can be determined by the exploration of the flow in this plane. The passage
of the vehicle leaves a trace or footprint in the wake, as a form of stagnation or total
pressure loss arising from viscous effects in the boundary layers and separated zones
and/or entropy jumps through shock waves. As the total pressure can be measured
with high accuracy, the overall accuracy in determining the drag from this method
is higher, especially in the case of low magnitude drag force. Figure 8.17 shows the
total pressure distribution downstream of a transport aircraft, a half-wing model in
a Mach 0.7 flow. This Figure first shows a total pressure loss region, linked to the
viscous drag from the boundary layers, and a second region corresponding to the
turbulence in the wingtip vortex which contributes to induced drag.

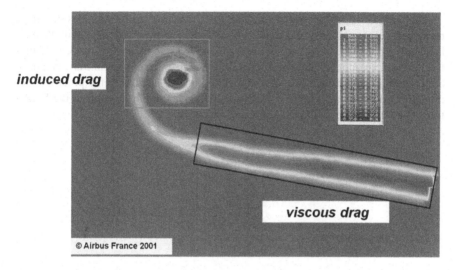

Fig. 8.17 Viscous wake and tip vortices downstream of a half-wing model at Mach 0.7

Fig. 8.18 Survey of the wake of a half model of Airbus in the S1MA wind tunnel (© ONERA)

This kind of survey is very instructive for aircraft or ground vehicles designers because it helps to identify and characterise the drag sources and to quantify them. It is a guide to improve the design of a wing by addressing the factors that contributes to viscous drag, induced drag and wave drag due to shock waves in transonic flow. Such an approach, called drag decomposition, is also applied to the processing of numerical simulations. Figure 8.18 shows a wake exploration setup in the S1MA transonic wind tunnel at ONERA, Modane-Avrieux.

Chapter 9
Characterisation of Flow Properties at the Surface

9.1 The Action of a Fluid at the Wall

The main forces acting locally on a surface experiencing a fluid flow are due to the pressure and the wall friction, the latter being responsible for heat transfer between the surface and the fluid. The direction of the shear stresses is of other practical importance. The pressure and wall shear stress or skin friction is of obvious importance in determining vehicle lift and drag in the absence of direct force measurement. However, the aerodynamic forces are very rarely determined from the integration of the local contribution as it would require extremely dense measurements of pressure and friction over the whole surface. The distribution of pressure on a vehicle is necessary to determine the local loads on the structure, and also provides an indication of the organisation of the flow locally, for instance a pressure plateau revealing a separated region. This is less true in three-dimensional flows where separation is a more complex process that cannot be inferred from the sole inspection of wall pressure distributions.

The knowledge of the skin friction is useful not only to determine the drag, but also to validate the theoretical models. This quantity is indeed difficult to predict with precision, since its calculation depends on both the precision of the numerical scheme (the friction being proportional to the derivatives of the velocity) and the robustness of the physical model. In addition, in two-dimensional planar or axisymmetric flows, the change in the sign of the shear stress at the wall is the best indicator of separation.

The ability of computer codes to predict the location of this point is a good indicator of their accuracy. In three-dimensional flows, the definition of separation is more subtle, as the transverse velocity component is not equal to zero and drives the line of separation in its direction, except in some particular situations.

Heat transfers (or flux) are of vital importance for flying at high Mach numbers (greater than 3), where some parts of the structure are subjected to considerable heating in hypersonics (re-entry body in particular). Also, high temperature flows

© Springer Nature Switzerland AG 2020
B. Chanetz et al., *Experimental Aerodynamics*,
Springer Tracts in Mechanical Engineering,
https://doi.org/10.1007/978-3-030-35562-3_9

are present in propulsive nozzles, in the base region of missiles or space launchers, to quote a few examples. Since it involves transfer phenomena, like skin friction, the heat flux is a delicate quantity to predict accurately and therefore a good candidate for the validation of the calculation codes. The heat transfer distributions can also reveal changes in the nature of the flow, for example the transition of a boundary layer from laminar to turbulent state.

In addition of resolving the physical quantities on the surface of the models, the measurement of their deformation increases the level of complexity during wind tunnel testing. Indeed, significant load causes non-negligible deformation of the structure. For example, in the wind tunnel the wing can be twisted by 1°, a non-negligible amount and compromises the comparison with a CFD calculation which can show significant variations. This increases the necessity of characterising the deformations so as to perform the simulation on the real shape of the model under the aerodynamic loads. This previously underestimated effect is now systematically evaluated.

9.2   Measurement of Pressure at the Wall

9.2.1   Pressure Scanning Systems

The measurement of the pressure on the surface of a model is conventionally carried out through small orifices or tappings (diameter of 0.3–0.5 mm) connected to a sensor or transducers via a tubing. This technique is well known and there is a wide variety of transducers based on the magnitude of pressure to be measured, the response time (case of unsteady measurements), the size and cost. Due to the large number of tappings on a model in order to have detailed information on the pressure distribution on its surface, several tappings are most often connected to a common transducer via a scanning device. The pressure transducer is installed in the scanner that is mechanically driven, the pressure ports of the scanner being successively in communication with the transducers. Figure 9.1 shows a ScanivalveTM scanner with 4 heads of 48 channels each. The scanner most often includes a calibration valve bringing the sensor into communication with known pressure prior to tests. The main disadvantages of these scanners are their size, often prohibiting their installation in the model and their relatively slow scanning speed (few readings per seconds).

The electronic scanner enabling a considerable increase in the scanning rate (up to 20,000 readings per second) consists of a set of transducers, each in communication with its own pressure port. The transducer is generally a Wheatstone bridge diffused in a monocrystalline silicon crystal by means of a semiconductor type treatment (see below). The voltage outputs of the transducers are sent to multiplexers and can be digitised selectively. Electronic scanners are quite miniaturised in the way to be imbedded in small scale models and positioned as close as possible to the tappings to reduce the response time. The number of sensors per scanner is generally between 20 and 50. Figure 9.2 shows a 16-channel electronic pressure scanner with reference and calibration.

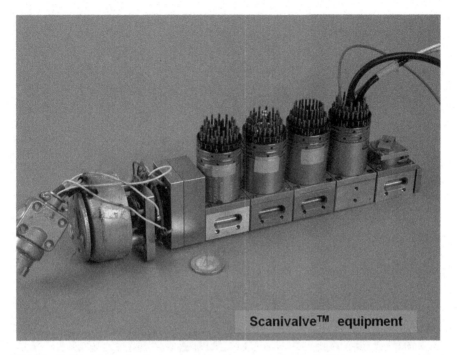

Fig. 9.1 48-channel four heads pneumatic pressure scanner Scanivalve™

Electronic scanners have virtually superseded pneumatic scanners. Figure 9.3 shows a 64-channel ScanivalveTM electronic dispenser and a pressure tap of the type used to measure the pressure on cars tested in the wind tunnel. While mounted on the vehicle, these taps avoid having to drill holes through the body and windows.

9.2.2 Types of Pressure Transducers

In the early days of wind tunnel testing, the pressure was generally measured by means of multi-tube manometers, by recording the height of the displaced liquid column. The measurement is made with respect to a reference pressure applied to the other end of the tube or the reservoir, the reference being atmospheric pressure or a vacuum. Depending on the range of pressure, the reservoirs and the tubes are filled with water, alcohol or mercury. Although very useful for monitoring the progress of an experiment, manometers of this type are hardly used for quantitative measurements.

Instead, transducers are used to translate the pressure on a sensing material into an electrical signal that can be amplified and digitalised. There are three categories of pressure sensors (see Fig. 9.4): the relative sensors that measure the pressure

Pressure Systems™ equipment

Fig. 9.2 Pressure Systems™ 16-channel electronic pressure scanner plus reference and calibration

with respect to a reference pressure, the differential sensors measuring a pressure difference and the absolute sensors measuring the pressure with respect to the vacuum.

There is a large range of pressure transducers which operate on different physical principles. In the so-called passive transducers, the pressure force acts on a membrane whose deformation is detected and measured via a physical process. In passive sensors with strain gauges, the stress induced by the pressure causes a variation of the resistance of the gauges glued on the membrane (see Fig. 9.5a). This variation of resistance is conventionally measured by the Wheatstone bridge technique (same method as that used in the force balances described in Sect. 8.2.1). In capacitive and variable reluctance transducers, the deformation of the membrane placed in a gap induces, in the first case a variation in capacitance, in the second a variation in reluctance (see Fig. 9.5b). These variations are detected and measured by appropriate electronic circuits.

In active transducers the action of the pressure is detected directly. Thus, for transducers with piezoresistive element, gauges are diffused in a silicon membrane (see Fig. 9.6a). The force resulting from the pressure on the membrane generates a potential difference by piezoelectric effect, also used in aerodynamic balances (see Sect. 8.2.4). In a second category of the piezoelectric type, the pressure force is applied directly to a piezosensitive element (see Fig. 9.6b). Figures 9.7 and 9.8 show examples of piezoresistive type transducers, including a miniature series.

Fig. 9.3 64 channels electronic scanner Scanivalve™ and pressure tap to be mounted on a wall

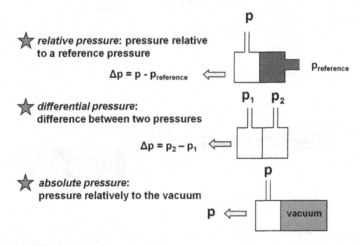

Fig. 9.4 Different types of pressure sensors

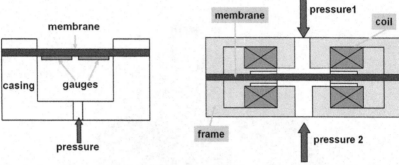

(a) Passive transducer with strain gauges   (b) Passive transducer with variable reluctance

Fig. 9.5 Passive pressure transducers

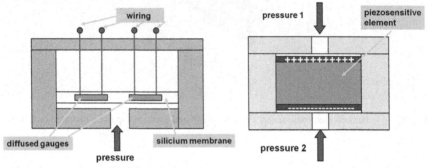

(a) Active transducer with diffused piezoresistive   (b) Active transducer with piezoelectric
element   element

Fig. 9.6 Active pressure transducers

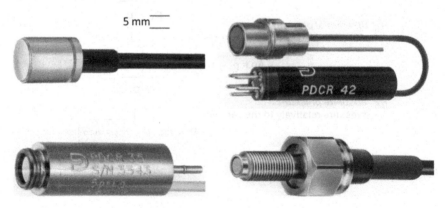

Fig. 9.7 Piezoresistive transducers of semi-conductor type diffused in a silica diaphragm

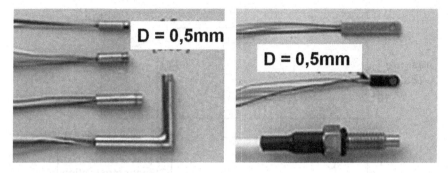

Fig. 9.8 Miniature pressure transducers

9.2.3 Sensitivity and Response Time

Depending on their type, the available transducers cover a very wide range of pressure from a few Pascals to several hundred bars for membrane transducers. Their response ranges from zero (stationary measurement) to approximately 10^5 Hz (see Fig. 9.9).

Whereas active pressure transducers have a measurement range extending from a few Pascals to several thousands of bars; their frequency range extends from zero (steady measurements) to almost 10^9 Hz (see Fig. 9.10).

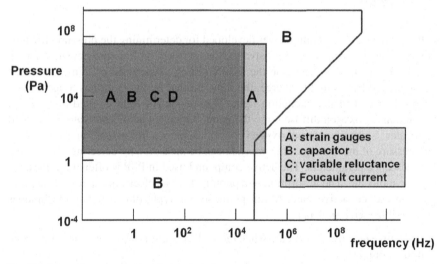

Fig. 9.9 Measuring range of membrane transducers

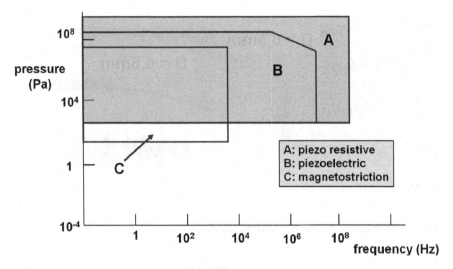

Fig. 9.10 Measurement range of active transducers

9.3   Pressure Sensitive Paint

9.3.1   *Principle and Composition of the Pressure Sensitive Paint*

In the early 1980s, a technique was developed for determining the pressure distribution on a model using a less intrusive technique based on Pressure Sensitive Paints (PSP). This method is based on the light emitting property of certain compounds when excited by a suitable source, the light emitted by the paint has longer wavelength than that of the excitation light. The amount of light emitted depends on the amount of oxygen diffused into the paint, which tends to deactivate the excited molecules. Since the internal oxygen concentration is a linear function of the external pressure of the same gas, the pressure can be measured by detecting some of the luminescence parameters. The active compound used in PSP is often a luminescent organic molecule. The absorption of a photon alters its electronic state, causing it to go to an excited, active state. At this point, several types of deactivation techniques are possible (see Fig. 9.11):

– a radiative deactivation occurs when the molecule emits a photon before returning to its ground state,
– the other non-radiative deactivation corresponds to the internal conversion of energy into heat,
– a third deactivation induces a transfer of energy by collision of the excited molecule with another oxygen molecule, bringing the molecule back to its ground state without emission of photons: this is commonly known as quenching.

The intensity of emitted light depends on the wavelength of the excitation light and its intensity, the concentration of active molecules, the thickness of the paint,

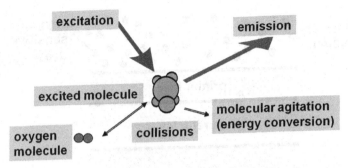

Fig. 9.11 Response to a light excitation of a molecule of the PSP active component

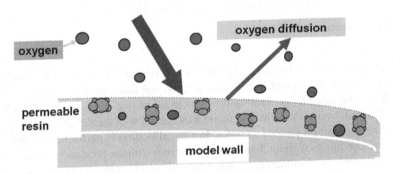

Fig. 9.12 Oxygen diffusion in the PSP coating

the surface temperature and the pressure. The temperature increases the irradiative deactivation rate and results in a lower intensity. The active molecules are embedded in a matrix controlling the diffusion of oxygen in the paint, silicone polymers being often used because of their high permeability (see Fig. 9.12).

The PSP coating, which additionally includes a screen layer and a tie-layer, must be smooth, uniform and adhere perfectly to the surface (see Fig. 9.13).

9.3.2 Relation Between the Pressure and the Luminous Intensity

The relation between pressure, p, and luminous intensity, I, is given by the Stern-Volmer equation:

$$\frac{I_0}{I} = A + B\frac{p}{p_o}$$

where I_0 is the luminous intensity emitted at a reference pressure, p_o, (condition without flow most often).

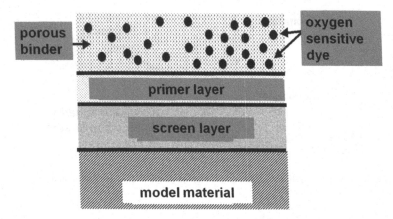

Fig. 9.13 Constitution of a PSP coating

The intensity ratio eliminates the influence of non-uniformity of both the illumination source and the paint layer (concentration and thickness). The coefficients A and B are evaluated by calibrating the paint as a function of pressure and temperature. This calibration is carried out in the wind tunnel with the model or in a separate pressure chamber on painted samples. In situ calibration is performed for different pressure conditions without flow or in comparison with the pressures recorded from pressure tappings. A difficulty in using the PSP is its simultaneous response to the change in temperature, which can be circumvented, either by using a paint that is not very sensitive to temperature, or by making a correction from the measurement of the wall temperature. Figure 9.14 shows an installation used for measurement of the pressure distribution on the central body of an aerospike launcher nozzle. It shows the optical fibre used to transmit the light illuminating the model and the CCD camera collecting the emitted light.

9.3.3   Main Areas of Application of the PSP

Pressure-sensitive paints are commonly used in the transonic regime where pressure variations are significant and temperature changes still modest (see Fig. 9.15). Figure 9.16 shows a model of a business jet's surface coated with PSP before testing in ONERA's S2MA wind tunnel.

In the supersonic regime, PSP measurements have been made on an aerospike nozzle with a PSP component having a low sensitivity to temperature (pyrene molecule for example). Figure 9.17 shows a false-colour representation of the pressure distribution measured by PSP on the central body of the nozzle. One can distinguish the traces of the jets of the primary nozzle impacting the central body as well as the traces of the transverse jet used for thrust vectoring (see also Fig. 7.5).

The sensitivity limit of the PSP can be reached at low subsonic speeds, the pressure variations being low (the PSP responds to the absolute value of the pressure).

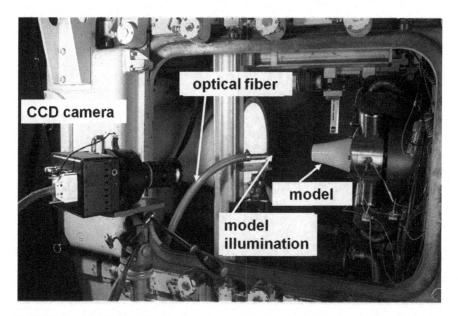

Fig. 9.14 Installation of a PSP system in the test section of the R2Ch wind tunnel at ONERA, Meudon (© ONERA)

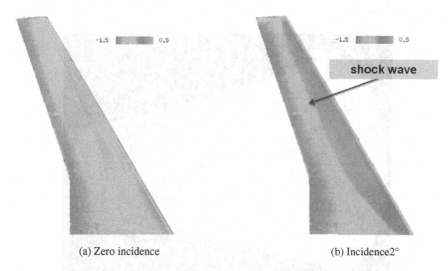

(a) Zero incidence   (b) Incidence2°

Fig. 9.15 Measurement of the pressure distribution on a swept wing in transonic flow (© ONERA)

Fig. 9.16 Business jet model prepared for PSP measurements in the S2MA wind tunnel at ONERA, Modane-Avrieux (© ONERA)

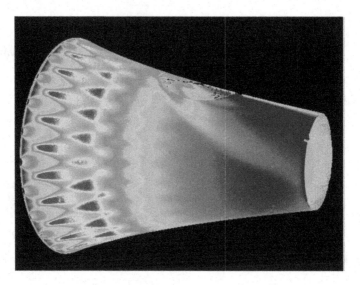

Fig. 9.17 False colour image of the pressure distribution on the central body of an aerospike nozzle (© ONERA)

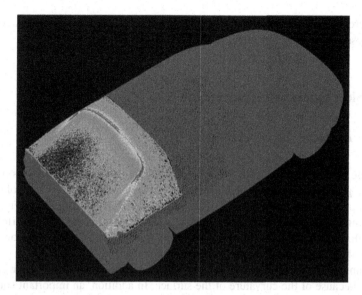

Fig. 9.18 PSP measurement of the pressure distribution on the rear part of a Peugeot 206 model (© ONERA)

However, if special precautions are taken in the calibration and in the processing steps, measurements at speeds as low as 30 m/s may be performed (see Fig. 9.18).

Being sensitive to the partial pressure of oxygen, the PSPs are in principle unusable in cryogenic wind tunnels where the gas used is nitrogen (see Sect. 4.3.4). The DLR, however, has developed a coating that can be used under cryogenic conditions, but that requires the introduction of a perfectly controlled small quantity of oxygen into the circuit of the wind tunnel.

9.3.4 Developments in the Field of PSP

Presently, PSP research focuses on the development of a dynamic version called UPSP for unsteady PSP. The porous binder in which the active molecule is inserted creates a response time of the order of one second, in order to increase the sensitivity sufficiently for unsteady measurements, it is necessary to remove this binder. There are two solutions for this: anodising and adding a porous ceramic. In the first case, an aluminium insert is anodised, which has the effect of creating micro-channels in which the sensitive molecule is deposited. This solution works well and allows measurements up to several kHz. The major disadvantage is the use of an aluminium insert, which involves special models. The other solution, which has been more studied as it has a similar structure as the conventional painting, consists of using a porous ceramic adsorbing the active compound. This method has a response time of the order of a millisecond, which is often sufficient. Its main disadvantage is the roughness which causes problem in maintaining a laminar flow.

9.4   Skin Friction Measurement

9.4.1   Floating Element Balance

Accurate measurement of skin friction is challenging especially in regions of high-pressure gradient. The direct method is to use a floating element consisting of a small insert mounted flush to the surface of the model, but isolated from it (see Fig. 9.19). The element is attached to a force balance which measures the force exerted on it by the flow. This is the most direct method for determining the skin friction as it is not based on any theoretical consideration of boundary layer properties. In practice, the technique of the floating element is difficult to implement on wing or aerofoil models because of the curvature of the surface. In addition, an important source of error can come from an imperfect alignment of the sensing element with the surface and also from the influence of a pressure gradient in the flow. The floating element method can be used to calibrate other techniques considering a flat plate situation for which this method can be very accurate.

9.4.2   Hot Film Surface Gauge

A more convenient technique is to use surface gauges formed by a heated platinum thin film (or wire) fixed to the surface of the model through an insulating insert.

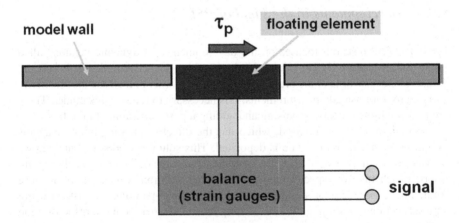

Fig. 9.19 Direct measurement of the surface shear stress: floating element inserted in the surface

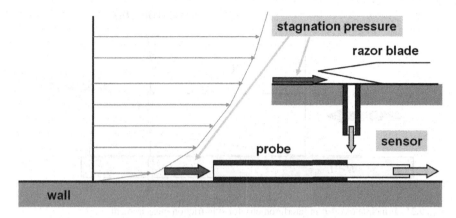

Fig. 9.20 Preston tube skin friction measurement

Measurement of the resistance of the film (or wire) makes it possible to deduce its temperature change under the action of the cooling or heating from the flow, the method being based on the Reynolds analogy between the heat flux and the skin friction. The device which must be calibrated is successfully implemented in supersonic and hypersonic flows. Its advantage is the small size of the sensitive element, which allows a local measurement, and a short response time (hence the use in hypersonic blow down facilities). In principle, it is possible to determine the skin friction in three-dimensional flows, where the shear stress at the wall is a two-component vector, by placing two gauges at a given angle on the surface of the model.

9.4.3 Stanton and Preston Tubes

The Preston tube technique consists of measuring the total pressure given by a Pitot tube placed in contact with the surface. In the Stanton tube technique, the detection part is sometimes made up of a razor blade glued to the surface with a static pressure port underneath (see Fig. 9.20). The skin friction is deduced from the recorded dynamic pressure based on the universal logarithmic law for the velocity distribution in the turbulent boundary layer in the near wall region (see Sect. 9.4.6). These devices often require calibration.

9.4.4 Oil Film Interferometry

This process is based on measuring the rate of deformation of a thin layer of oil deposited on the surface of the model, as shown in Fig. 9.21, the shear stress at the wall being deduced from the lubrication theory. If the film of oil is thin in comparison to its length, its surface then takes the form of a small wedge which has low intrusive effects.

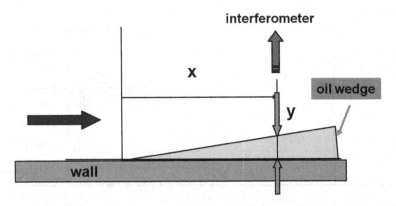

Fig. 9.21 Principle of oil-film interferometry for skin friction measurement

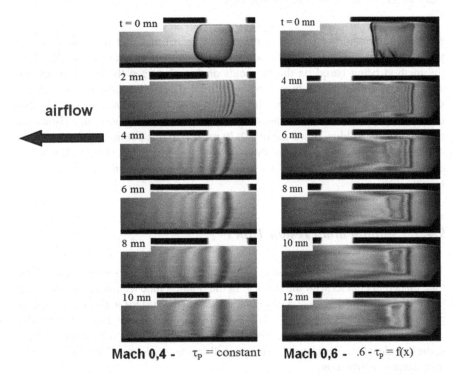

Fig. 9.22 Oil film method. Measurement of film thickness by interferometry (© ONERA)

The thickness y at any instant t can be measured accurately by an interferometry technique (see Fig. 9.22) and given as

$$y = \frac{\mu\, x}{\tau\, t}$$

where μ, is the dynamic viscosity of the oil, τ, the skin friction, and x, the distance from the tip of the wedge. The main issue with this technique lies in the fact that the oil film shears accordingly with the transient behaviour of the wind tunnel, from start-up till test speed and making difficult to identity initial time corresponding to a particular speed measurement. This introduces challenges in locating the apex of the oil film.

These challenges could be addressed by measuring the thickness of the oil wedge at two closely spaced points, which makes it possible to determine the effective "time" and "initial" time parameters, hence the skin friction. The method can be extended to three-dimensional flows if the direction of the skin friction vector at the wall is known.

9.4.5 Liquid Crystals Thermography

The skin friction can be determined using liquid crystal thermography, a combination of optically active organic compounds with the property of reflecting light at a particular wavelength in response to stimuli such as temperature, pressure, shear stress, magnetic and electric fields. For wind tunnel applications, it is possible to retain only one stimulus by suppressing the response to others. Thus, there are compounds that respond only to shear stresses under certain circumstances (or temperature for heat transfer measurements, see below). In practice, the model is covered with a thin film of the substance and illuminated by a white light source, the liquid crystal film then producing an image in the visible range which is recorded by a camera. The technique has several advantages: its very high sensitivity, its reversible nature and the ability to provide information on an entire surface. Thus, the liquid crystals are used to identify the laminar to turbulent boundary-layer transition region, the transition process being visualised by a colour change. Quantitative measurement of the skin friction is in principle possible by determining the colour changes under the action of the shear stress. An image analysis by digital image processing allows the determination of the wavelength λ of the reflected light. The local value of the skin friction is then deduced from a calibration giving the wall shear stress τ as a function of λ.

9.4.6 Adjustment Based on the Logarithmic Law
of Turbulent Boundary Layers

At sufficiently high Reynolds numbers, the turbulent boundary layer adopts an equilibrium state characterised by a velocity distribution satisfying the so-called well-defined law of the wall for turbulent boundary layers shown in Fig. 9.23. Starting

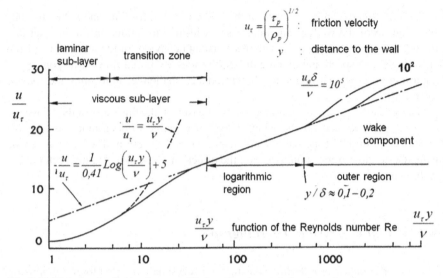

Fig. 9.23 Velocity distribution in an equilibrium flat plate turbulent boundary layer

from the wall, in the viscous sub-layer the flow is dominated by the viscosity, following the buffer zone, the universal logarithmic region is governed by the inertial character of the flow, finally at the outer boundary the wake component.

In the logarithmic region, the velocity distribution obeys a relation of the form:

$$\frac{u}{u_\tau} = \frac{1}{0.41}\mathrm{Log}\frac{u_\tau y}{v} + 5$$

where $u_\tau = \sqrt{\frac{\tau_p}{\rho_p}}$ is the friction velocity, τ_p, the shear stress at the wall, ρ_p, the density in the vicinity of the wall, y, the distance normal to the wall and, v the kinematic viscosity.

This method of determining the skin friction is based on resolving the boundary layer profile as close as possible to the wall then to re-plot the relation above by estimating a value of u_τ (trial and error), that is to say of the skin friction coefficient:

$$C_f = \frac{2\,\tau_p}{\rho_e\,V_e^2}$$

where ρ_e and V_e are the density and the speed at the boundary layer edge respectively. The correct value is obtained when there is agreement between the measurements and the universal logarithmic law. The method is reliable and accurate, provided that the boundary layer satisfies well-established equilibrium conditions, which is not the case if it is subjected to a pressure gradient, or if the Reynolds number is sufficiently low. The technique is also referred as the Clauser Chart technique.

Skin friction can be also determined from boundary layer probing by fitting the measured profile to other available theoretical laws, such as Coles' law. The method consists in iterating on the guessed values shear stress until the best overall agreement between the theoretical and measured profiles is obtained. If the point-to-point adjustment of the velocity distribution is too delicate, it could be based on the shape parameters adjustments. This second method is interesting if the probes cannot be positioned close enough to the wall to obtain a good definition of the logarithmic part of the profile. It also applies to boundary layers subjected to pressure gradients.

9.5 Measurement of the Wall Heat Transfer

9.5.1 Calorimetric Techniques

The local convective heat transfer between the flow and the model is most often determined using calorimetric techniques consisting of the measurement of the rate of change of the local temperature of the surface or of a sensing element (transducer) inserted into the surface. An inverse solution to the equation governing the thermal conduction through the wall of the model (or transducer) gives the heat transferred to the surface by the flow (energy per unit time and per unit area). In practice simplified forms of the heat equation are used by considering extreme situations leading to simple analytical solutions. If the heat flow can be assumed to be one-dimensional and if the wall is considered semi-infinite in thickness (so-called thick wall technique) then the heat equation is reduced to:

$$q(t) = \sqrt{\frac{\rho_m c_m \lambda}{\pi}} \int_0^t \frac{\frac{d\Delta T(\tau)}{d\tau}}{\sqrt{t - \tau}} d\tau \qquad (9.1)$$

where $q(t)$ is the rate of heat transfer (energy/unit of time/unit area), ρ_m, the density of the material constituting the wall, c_m its specific heat and, λ, the thermal conductivity. If the heat flow can be assumed constant over the duration of the measurement, the relation can be further simplified:

$$q(t) = \sqrt{\frac{\pi \, \rho_m c_m \lambda}{2}} \frac{\Delta T(t)}{\sqrt{t}} \qquad (9.2)$$

If the wall can be considered as infinitely thin (so-called thin skin technique), the heat flux is given by:

$$q(t) = \rho_m c_m \, e \frac{d\Delta T(t)}{dt} \qquad (9.3)$$

where e is the wall thickness. Here, the wall of the model must have a thickness of a few tenths of a millimetre, which prohibits its extension on models with complex shape.

Given the properties of the material constituting the wall, the heat flux is calculated by the Eqs. (9.1), (9.2) or (9.3) from a time recording of the wall temperature. The above methods require the rapid establishment of the flow so that the heat flux is applied almost instantaneously. This condition is achieved either by working in a fast-starting wind tunnel (blow down type), or by inserting the model into the flow after the nominal conditions have been established.

The simplifying assumptions underlying the formulae above are not always verified: the local curvature of the wall can be pronounced, the transient nature of thermal transfer, the multi-directional conduction (the lateral conduction), the wall material properties may vary with temperature and the wall thickness variation. In these circumstances, correction terms must be added and a more rigorous processing technique consisting of solving the heat equation by a numerical method is to be employed.

The most common method for measuring wall temperature is the use of thermocouples mounted into the model wall (see Fig. 9.24a). Surface films (platinum films) similar to those used to determine skin friction (see Fig. 9.24b) can be also employed. The films are most often deposited on a thermally insulating insert to minimise lateral conduction losses (see Fig. 9.25).

The above techniques are not valid if the flow and the model are in thermal equilibrium, which is often the case in transonic and/or supersonic continuous wind tunnels. The calorimetric method must then be modified by inserting a heat source (an electrical resistor) into the wall of the model near the hot film. Convective heat transfer due to the flow is deduced from the energy required to maintain the surface temperature constant.

Conventional calorimetric techniques have reached a high degree of sophistication and are still employed reliably even under extreme conditions. Their main drawback is their localised measurement, a good spatial resolution requiring the installation of

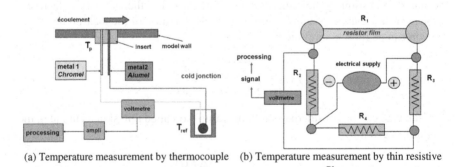

(a) Temperature measurement by thermocouple   (b) Temperature measurement by thin resistive film

Fig. 9.24 Determination of heat transfer by calorimetric technique

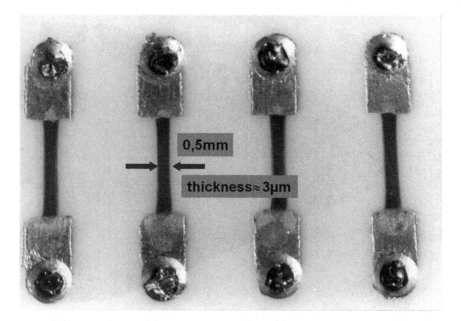

Fig. 9.25 Platinum films for measuring surface temperature

a large number of gauges or thermocouples to determine the distribution of heat flux on a complex shape (several hundred can be installed on certain hypersonic models).

Methods such as temperature sensitive paint (TSP) or infrared thermography (IR) do not have this disadvantage because they capture the distribution of heat flux over a larger portion of the model and they are almost un-intrusive when surface heating is not required. The IR-thermography has been significantly developed and takes advantage of the spectacular developments of infrared cameras, to the point that the accuracy obtained now outperforms that of local sensors. IR-thermography is becoming the reference method as sensors are only used in areas that are not optically accessible.

9.5.2 Thermo-Sensitive Paints

In thermo-sensitive paint, the model made of a thermally insulating material, is coated with a paint and depending on the temperature its colour ranges from green to blue, yellow, brown and black as the temperature increases, with colour changes occurring at well-defined temperature levels (see Fig. 9.26). The heat flux is determined by recording the variation of the colour during the test. A frame-by-frame analysis of the live recording determines the temperature history on the surface of the model by differentiating between the instants of the colour change.

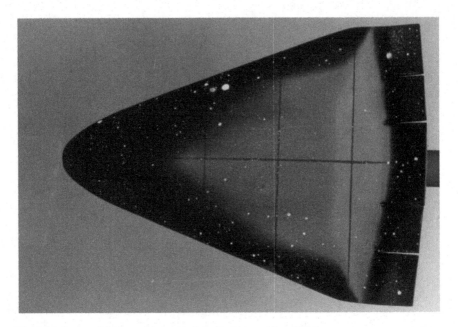

Fig. 9.26 Image of the heated lower surface of a hypersonic vehicle

By applying the appropriate heat conduction relations, the local heating based on the colour changes can be determined. A calibration can be also used by comparing colour changes with those on a sphere, for which the distribution of heat transfer is well known. This technique gives the distribution of heat transfer on a complex model from a single test, without having to implement more expensive equipment. Its main drawback is its irreversible nature, which limits its reuse the on model immediately.

This procedure is gaining more interest because of the parallel development with PSPs that are also temperature sensitive (see Sect. 9.3). The method is then called TSP for Temperature Sensitive Paint. It is thus possible to combine in the same coating, a component that is substantially insensitive to the temperature, which provides the pressure and a component that is more sensitive to the temperature, hence the temperature distribution. The same processing technique as for the PSP is applied making the process reversible. The advantage of this method over the IR is greater flexibility since ordinary cameras and optics can be used. But the IR is still fairly more advanced because of its high accuracy and the wide range of temperature it can handle, while the TSP is limited to 100 °C in practice.

As mentioned above, the liquid crystals also respond to the temperature by changing colour, hence the possibility of determining the heat flux by recording the temperature history of the model. The non-reversible nature of conventional thermo-sensitive paint is thus overcome, on the other hand this technique is more complex.

9.5.3 Infrared Thermography

All bodies emit radiation whose intensity is a function of the temperature T, distinction has to be made between the total intensity radiated over the entire spectrum and that radiated at a particular wavelength λ. For the black body, the radiated energy is given by Planck's law which states that the energy per unit of time (power) radiated in a certain direction OX by a unit surface of the black body—or radiative flux—is given by:

$$L_\lambda^0 = \frac{2h\,c^2\lambda^{-5}}{\exp\left(\frac{hc}{k\lambda T}\right) - 1}$$

where h is the Planck constant, k is the Boltzmann constant, and c is the speed of light. The radiative property of a real body is characterised by its emissivity defined as the ratio between the intensity of the light emitted by this body and the intensity of the light emitted by the black body at the same temperature.

In infrared thermography (IR), the model is observed by an infrared camera containing a detector element sensible to infrared radiation at a certain wavelength (the band 3–5 μm is the most used). Since the signal delivered by the camera is proportional to the radiative flux, this information must be converted into temperature in order to build the temperature map of the model. This can be done by calibrating the system by observing a black body whose temperature and emissivity are known. The calibration can also be done by placing thermocouples on the model that provide the temperature at selected points. The test section must be equipped with windows that allow infrared radiation, such as germanium for the 8–12 μm band, silicon for the 3–5 μm band or zinc sulphide. This becomes a serious constraint for infrared thermography.

Figure 9.27 shows an infrared camera mounted outside the test section of a supersonic wind tunnel. A thermal image of the model is obtained where the regions at different temperatures are represented by grey or false colour scales. This qualitative aspect of thermography is very useful for detecting the laminar to turbulent transition (see below and Sect. 9.1.7). By processing a series of images taken at known time intervals, it is possible to build the temporal history of the temperature of the model and to deduce the distribution of the heat flux on its surface from the heat equation. Figure 9.28 shows the distribution of heat flux on a hemisphere in a flow at Mach 5, in false colours, the hottest regions being indicated by "cold" hues, and vice versa.

The method is very powerful because it gives a complete picture of the distribution of heat flux over the model. In addition, the process is very sensitive and reversible, the model does not have to be changed or modified between each test. However, as seen above, the method requires a window which allows infrared radiation (unless the camera can be installed inside the test section). This is an important constraint, the choice in optics being limited. A minor disadvantage is the reduced size of the sensor: 640×512 pixels, which is small compared to what is achieved in the visible

Fig. 9.27 FLIR™ infrared camera in position for the study of transition in the ONERA S8Ch wind tunnel. Note the infrared window (© ONERA)

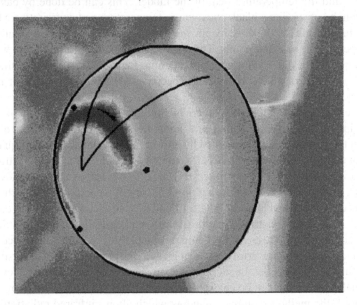

Fig. 9.28 False colour image of the heat transfer distribution on a hemisphere at Mach 5 (© ONERA)

range. Despite these limitations, infrared thermography tends to be systematically used for heat transfer measurements because of their versatility and ability to provide heat flux distribution over a body of complex geometry.

The infrared technique, which has a temperature resolution of a fraction of a degree, is widely used to detect the laminar to turbulent transition region on a model, making it a valuable tool for laminar flow research. Figure 9.29 shows a setup for detecting the laminar to turbulent transition on a flat plate at Mach 1.6 where infrared thermography and TSP were used in parallel. The IR and TSP images are compared in Fig. 9.30, the flow coming from the left. IR detection is also used for transition studies during flight tests (see Sect. 1.7.1).

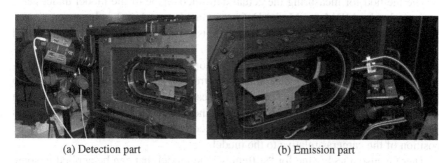

(a) Detection part   (b) Emission part

Fig. 9.29 Test set-up for the detection of the laminar to turbulent transition on a flat plate at Mach 1.6 by using TSP (© ONERA)

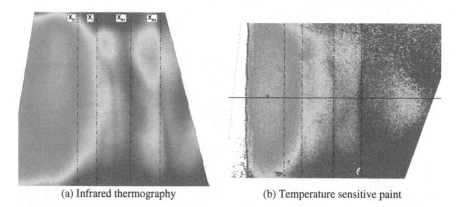

(a) Infrared thermography   (b) Temperature sensitive paint

Fig. 9.30 Detection of the laminar to turbulent transition on a flat plate at Mach 1.6. Laminar regions are in green (© ONERA)

9.6   Measurement of Deformations of the Models

One of the aims of testing in large wind tunnels is to validate performance pre-calculated by CFD simulation. The differences sometimes observed can be explained by various causes, among which are the interactions with the supports of the model and the effects due to the confinement of the wind tunnel walls (see Sect. 2.5). Another important cause is the deformation of the models under the influence of aerodynamic forces, in particular that of the wings. In case of considerable loading, the twist of a wing can reach 1° which is non-negligible. The comparison with simulation is then flawed due to the disparity with the deformed geometry during the experiment.

The method for measuring the actual deformed shape of the model under aerodynamic loads (MDM for Model Deformation Measurement) is based on stereo-imaging. As shown in Fig. 9.31a, the principle is to use two cameras that are focussed on the wing model.

The wing is equipped with fixed markers consisting of white "stickers" of negligible thickness, with a black disk at their centre (see Fig. 9.31b). By capturing two images using the stereo-vision it is possible to calculate the position of the markers in space and thus to measure the deformation of the model during the test. The cameras must be calibrated with a known pattern, to correct for the optical errors and the position of the cameras relative to the model.

There is also a technique for "stiffening" the model that can be applied in some wind tunnels. This technique consists in carrying out two consecutive tests, under the same Mach number conditions, but with different stagnation pressures. The measurements of the corresponding aerodynamic forces can then be linearly extrapolated to a zero-stagnation pressure for which the model is theoretically not deformed.

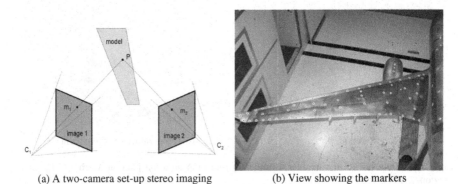

(a) A two-camera set-up stereo imaging   (b) View showing the markers

Fig. 9.31 Setup for measuring the deformation of a model subjected to aerodynamic loading (© ONERA)

Chapter 10
Intrusive Measurement Techniques

10.1 Solid Probes: Benefits and Limitations

Even nowadays the flow is still diagnosed by solid probes such as Pitot probes, thermocouples and hot wires in particular, it has been crucial in not only capturing the mean flow, but also in providing further insight into coherent structures. Despite more sophisticated measurement techniques, in many cases their use is justified by their accuracy, reliability, cost and simplicity. The main drawback are their intrusive effects, leading to scepticism during the measurement of shear layer developed by a separated flow and are more delicate in transonic flow. Besides, to avoid vibration of the probe the struts and mounts have the tendency to be voluminous and this introduces solid blockage which has an impact on the desired pressure distribution. Also, during spatial characterisation of the flow, complex and expensive traverse system is usually required and the displaced position of the strut could again impair with the local and overall pressure gradient. During the measurement in low pressure turbines and hypersonic flows, the probes need to be able to withstand very high temperature without disintegrating or deforming due to thermal expansions. An operational limit is usually set at approximately 1200 K and above this temperature more exotic material or sometimes cooling systems are required, which increases cost and complexity of the technique. In confined experimental facilities where reduced scale, miniature models are tested, deformation and vibration of the probes introduces large uncertainty in terms of positioning of the probe and hence compromising significantly the end result.

© Springer Nature Switzerland AG 2020
B. Chanetz et al., *Experimental Aerodynamics*,
Springer Tracts in Mechanical Engineering,
https://doi.org/10.1007/978-3-030-35562-3_10

10.2   Pressure Probes

10.2.1   Measurement of Stagnation or Total Pressure, Pitot Probe

The stagnation or total pressure of a flow can be measured using a Pitot probe which is basically a tube (hypodermic tube) with the opening or orifice facing the flow, as shown in Fig. 10.1. The resulting pressure is registered using a transducer connected to the Pitot probe through pipes.

This pressure corresponds to the loss in kinetic energy at isentropic conditions while the flow decelerates from maximum to zero at the tip of the probe. It differs from the absolute pressure which is the sum of the gauge and atmospheric pressure. When the fluid is at rest, at infinity the total pressure is equal to the atmospheric pressure which is sometimes referred as the reservoir pressure, after the fluid reservoir in traditional manometers. At subsonic conditions the stagnation pressure is equivalent to the gauge pressure which is measured relative to a vacuum and similarly for inviscid and attached flow. In supersonic flow the probe measures the total pressure downstream of the shock generated in front of the probe and using isentropic relation the local Mach number can be determined.

The frontal part of the probe could be flattened, as shown in Fig. 10.1b, in order to be positioned closer to the surface of the model. This is essential for capturing the velocity profile of boundary layers, mainly during the turbulent state as the profile is shallower and in order to resolve the universal logarithmic region accurately. Usually a near wall measurement of 0.2 mm (as shown in Fig. 10.2) is comfortably achieved. Flattened Pitot probes could be also used to capture the thin shear layer of jets and also in two mixing layers with different properties.

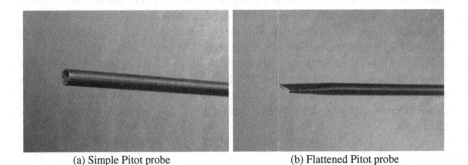

(a) Simple Pitot probe   (b) Flattened Pitot probe

Fig. 10.1   Examples of Pitot probes

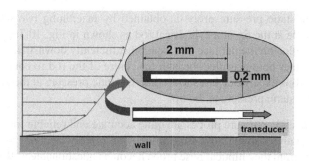

Fig. 10.2 A flattened Pitot probe for the measurement of boundary layers

10.2.2 Measurement Using Pitot-Static or Prandtl Probe

Measurement of the static pressure of a flow is not trivial, since the pressure is an intrinsic and local property of the fluid which cannot be determined directly, unless using the procedures described in Chap. 12. The only way of measuring the static pressure using a solid probe is by a pressure port drilled normal to the surface of the probe. Two types of static pressure probes are commonly used:

- The classic probes are normally made of an extended cylindrical tube with an elliptical or sharp conical nose, for subsonic and supersonic flow respectively. The pressure port is located in the region where the surface is uniform, sufficiently downstream of the leading tip (approximately 10 diameters), where the static pressure has equalised with that immediately upstream of the leading tip. More detail is shown in Fig. 10.3a where the probe contains four ports to minimise the effect of incidence and yaw.

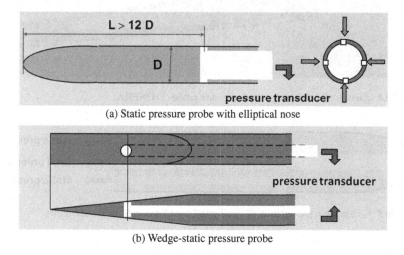

(a) Static pressure probe with elliptical nose

(b) Wedge-static pressure probe

Fig. 10.3 Schematic representation of two types of static pressure probes

– The wedge-static pressure probe is obtained by machining two flat faces at a shallow angle at the tip of a cylindrical rod as shown in Fig. 10.3b. The pressure ports are drilled on the flat face at a location sufficiently downstream of the sharp tip and another hole is drilled through the centre of the rod to connect the static ports. Previous experience has shown that the static pressure at the port is almost equal to the static pressure at the tip.

Both types of static pressure probes are quite accurate and reliable for the measurement of gradually evolving flow. But in flow with rapidly changing static pressure measurement from the elliptical nose probe becomes questionable as the pressure at the tip might be significantly different to that at the location of the port. In this case the wedge probe shown in Fig. 10.4 performs better.

From Bernoulli's principle the speed of an incompressible flow could be determine by the difference between of the total or stagnation pressure from a Pitot probe and the static pressure from a static pressure probe, this resulting pressure is also known as the dynamic pressure. This is quite common for measuring the speed of the flow in the working section of a wind tunnel. Alternatively, a Pitot-static or Prandtl probe shown in Fig. 10.5 is used to measure both the total and static pressures on the same probe.

In a supersonic flow the Pitot probe measures the stagnation pressure downstream of the bow shock formed ahead of the probe tip. Using this pressure the local Mach number can be deduced using normal shock theory. The local Mach number can be determined using a Pitot-static probe but here the isentropic theory is applied for compressible flow.

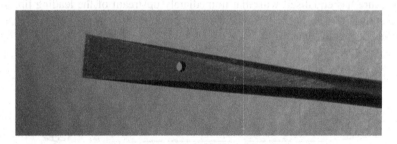

Fig. 10.4 An example of a wedge-static pressure probe (© ONERA)

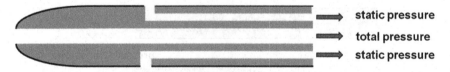

Fig. 10.5 Schematic representation of an elliptical nose Pitot-static or Prandtl probe

10.2.3 Multi-holes Probe to Resolve the Direction of the Velocity Vector

Pressure probes are also used to resolve velocity vectors in three-dimensional flows. Here, multi-hole probes are used and the most basic arrangement is a three-hole probe shown in Fig. 10.6. The middle hole behaves like a Pitot probe and two side probes with the tips machined at angle and placed symmetrically about the middle probe resolves the direction of the flow. Again, the probe could be flattened so as to reduce its thickness and could be also bent in such a way that it is resting on the surface. This probe is commonly used to traverse boundary layers where the plane parallel to the surface constitutes of two velocity components.

A minimum thickness of approximately 0.2 mm can be achieved and a width of approximately 1.5 mm. The tip of the probe must be aligned in such a way that it is in the plane parallel to the surface. The magnitude of the velocity is deduced from the total pressure captured by the centre Pitot probe and, a local static pressure tapping

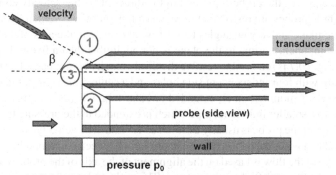

(a) Schematic of the measurement setup using a three-hole probe

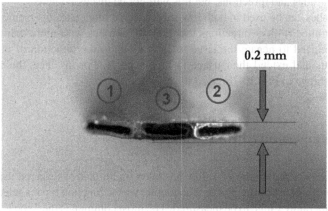

(b) Probe seen from the front

Fig. 10.6 A flattened three-hole probe for boundary layer measurement

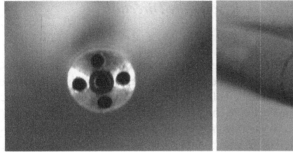
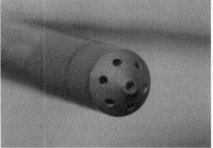

(a) Five-hole probe with a diameter of 1.5 mm   (b) Seven-hole probe with a diameter of 1.5 mm

Fig. 10.7 Multi-hole probes for the measurement of three velocity components (© ONERA)

mounted flush to the surface; the direction is resolved from the difference between the slanted side tubes. The most accurate procedure is to rotate the probe until the pressure difference between side tubes is almost zero or very small and at this yaw position the axis of the probe is aligned in the direction of the velocity component.

Three-hole probes are restricted to the measurement of two velocity components in the same plane and at yaw angles less 40° with respect to the axis of the probe. If the third component normal to the plane is not negligible then a five-hole probe, as shown in Fig. 10.7 is required to resolve the third velocity component. The probe has a conical tip with four independent side holes and a centre hole right at the nose. The conical probe is mounted on an extended circular tube where each pressure port is connected to a smaller individual pipe which is extended to the pressure transducers. The diameter of the probe is usually as small as 1.5 mm. The velocity components are determined by taking the difference in the pressure between corresponding ports and the direction of the flow is based on the alignment of the axis of the probe with respect to the tunnel or line-of-flight axis systems. The multi-hole probes are calibrated for both magnitudes and directions, but this will not be discussed here.

The probes can be used in both low subsonic and moderate supersonic flows, but they are limited to flow directions less that 40°. Probes with more than five holes were introduced to negate the issue of limitations to the direction of the velocity components and extend their usage to higher angle. For instance, the seven-hole probe, as shown in Fig. 10.7b.

10.3   Temperature Probes

The stagnation temperature of a flow has been traditionally measured using a temperature probe constituting of a thermocouple, whose junction type is selected based on the temperature range to be measured. The most common is the Type K (chromel-alumel) and in its most primitive form it could be used by just placing the bare sensing element in the flow, as shown in Fig. 10.8a. These probes are fairly small

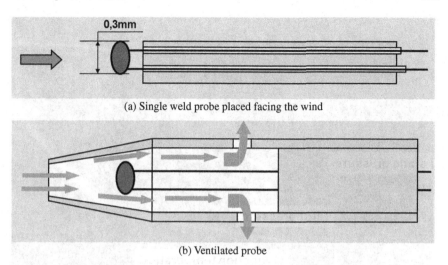

(a) Single weld probe placed facing the wind

(b) Ventilated probe

Fig. 10.8 Stagnation temperature probes

in size and at the same time they have a reasonable frequency response for reliable measurements. However, the frequency response is a function of the Mach number and therefore a calibration is required for measurement in compressible flow. The frequency response can be improved by encasing the thermocouple junction in a tube with a faired tip as shown in Fig. 10.8b; inside the tube the air circulates at lower speed due to the side breathing holes downstream. This type of ventilated probe performs better than the bare probe, but it tend to be much larger in size.

The response of the thermocouple's junction is directly influenced by its shape and size. For a faster response time the junction's size is reduced. But this could be an issue for measurement in blow-down tunnel, where the sensing element could experience losses due to thermal convection effect within the flow, conduction from the support prongs and radiation. The effects vary from region to region in the flow and this is usually not accounted in the calibration curve. Figure 10.9 shows a probe with the capability of measuring the total pressure and temperature and the static pressure simultaneously.

10.4 Measurement Using Hot Wire and Hot Film Anemometry

10.4.1 Basic Principles

Hot wire anemometry is a classic and well-established technique for measuring the velocity of airflow. The principle of operation is based on the heat transfer between

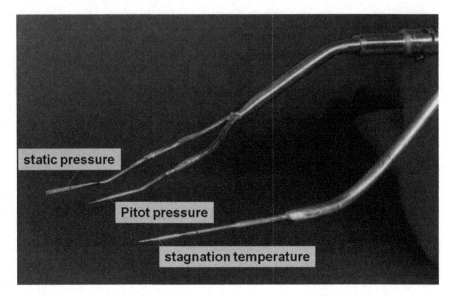

Fig. 10.9 A probe for measuring total pressure and temperature and static pressure altogether (©
ONERA)

the micrometre size wire or film heated by an electric current and the airflow passing
over it. The cooling effect from the airflow changes the temperature of the wire and
hence its resistance or the electrical energy required to keep the resistance and hence
the temperature constant. These quantities can be measured accurately by placing the
hot wire in a Wheatstone bridge arrangement. Following a calibration, the relation
between the voltage and the resistance of the wire or between the voltage and the
flow speed can be derived. The calibration has to be undertaken in conditions similar
to that of the experiment and sometimes in situ calibration is recommended. The
feedback loop in the electronic circuit should be adjusted for an optimum frequency
response for a particular flow, while keeping the electronic noise as low as possible.
Post-filtering can be applied to allow the passage of only a certain frequency band,
most relevant to the flow. Hot film sensors are made of a thin film of platinum or
nickel deposited on an isolated substrate, usually quartz. They are less fragile than
hot wires which are usually made of 2.5 or 5 μm tungsten wire and are mainly used
for liquid flows or highly contaminated flows. It has to be noted that their response
time is a lot lower than hot wires.

The hot wire is very sensitive to the velocity, density and total temperature of
the flow. For incompressible flows at Mach number less than 0.3, by definition the
density is constant and therefore the hot wire captures the velocity directly. The
result is more difficult to interpret in compressible flows where the density changes.
Here, it is needed to distinguish whether the hot wire is responding to the change in
the speed of the flow (vorticity mode), the pressure (acoustic modes) or temperature
changes (entropic mode). There are measurement techniques available to separate

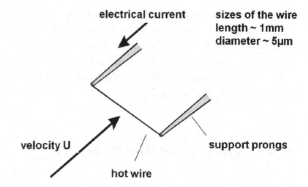

Fig. 10.10 A single wire probe for the measurement of the normal velocity component

the influence from each of these three modes and this usually involves varying the temperature of the wire. In practice, hot wires are mainly dedicated for the measurement of velocity fluctuations and turbulence at subsonic speed, but their application has been extended for the measurement in the transonic and supersonic regime, at a relatively high cost and complexity for calibration and data analysis.

As shown in Fig. 10.10, the hot wire is simply made of a very thin wire welded to the tip of two support prongs. The sensor wire is usually of a diameter of 5 μm and a length of 1 mm. In some cases, a shorter probe is used for better spatial resolution and to keep the length to diameter ratio (important parameter) high enough the diameter is usually reduced to 2.5 μm. This size of wire is also more commonly used for high speed flows. The biggest advantage of hot-wire is their high frequency response which makes them ideal for measuring unsteady flows at very high frequencies, of the order of several hundred of kilohertz.

10.4.2 Modes of Hot-Wire Anemometry

Depending on the types of measurement and personal preferences three different modes can be adopted:

– In the constant current anemometry (CCA) mode, the current circulating through the wire is kept constant and the voltage change due to the change in the resistance of the wire, from the cooling effect of the flow is measured.
– In the constant temperature anemometry (CTA) mode, the resistance of the wire is kept constant and through a feedback loop the current across the Wheatstone bridge is adjusted.
– In the constant voltage anemometry (CVA) mode (not very common), the voltage through the wire is kept constant and therefore the variation in the current and the resistance in the loop is measured.

From the signal of the hot wire both the mean and the fluctuating components of the flow can be determined. Further analogue and/or digital treatment of the signal allows the estimation of statistical quantities such as RMS values, skewness and kurtosis and from further analysis spatial and temporal correlations of the flow can be derived. Due to the fast frequency response and high signal to noise ratio of the hot wire it is invariably the most suitable candidate for capturing frequency spectra, where the corresponding flow mechanisms and the more importantly the turbulence energy cascade can be resolved.

10.4.3 Types of Hot Wire Probes

Figure 10.11 shows different types of probes for hot wire anemometry applications.

For the measurements of multiple velocity components two or more wires are required. The most common is the X-probe shown at the bottom of Figs. 10.11 and 10.12 which can measure the mean longitudinal and transverse velocity compo-

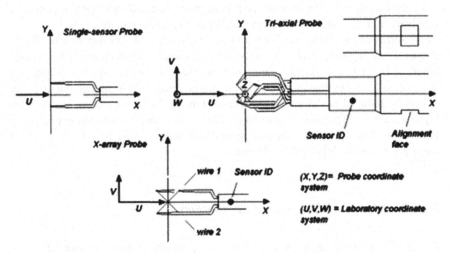

Fig. 10.11   Types of probes used for hot wire anemometry (© DANTEC)

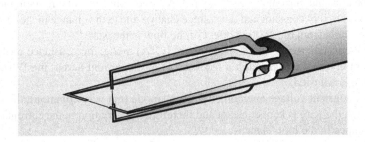

Fig. 10.12   A two wires X-probe for the measurement of two velocity components (© DANTEC)

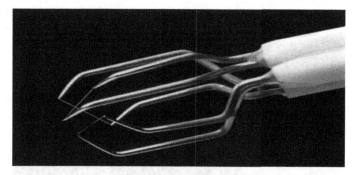

Fig. 10.13 A tri-axial probe for measurements in three dimensional flows (© DANTEC)

nents, U and V respectively. The RMS value can be used to determine the normal Reynolds stresses and by applying some correlations the Reynolds shear stresses can be resolved.

For the measurement of three components a three wire X-probe can be used as shown in Fig. 10.13 and provides the complete velocity vector including all the Reynolds stresses. Another method is to use a single yawed or slanted probe and to rotate it accordingly; however, the accuracy in resolving the Reynolds stresses is questionable due to temporal variations in the flow.

10.4.4 Applications

Hot wire probes are mainly used to capture velocity fluctuations at low speeds for the study of laminar, transitional or turbulent flows. Their use was extended to supersonic and hypersonic flows where the effect of compressibility introduces very complex signal analysis. But this complex signal analysis can be beneficial to identify the various other modes (vorticity, acoustic and entropic) that the wire will respond to. In high speed flows it is often necessary to retract the probe in and out of the flow before and after measurement to avoid parasitic effects due to vibration. At a Mach number greater than 1.2 the length scale of the bow shock due to the wire is assumed to be close to the mean free path of a molecule, so continuum applies. However, transonic flows are more complex to deal with so the calibration process is more meticulous.

In separated flows hot wires are not able to cope with the recirculating components due to their directional insensitivity. A technique called flying wire could be employed but the complexity of the traverse mechanism increases significantly and the effect of the prongs and probe supports in the recirculation region is debatable. For this particular reason optical techniques have seen an extended use however they cannot compete with the frequency response of the hot wire which is important for measurement of transitional and turbulent flows. Due to their microscopic size the hot

Fig. 10.14 An assembly of rakes consisting of 143 hot wires probes during the boundary layer measurement in the WALLTURB facility (© Institut PPRIME)

wire probes are susceptible to frequent damage, however they are relatively simple to operate and cheap. The issues related to phase-shits or lags in a spatially evolving flow can be addressed by employing multiple probes on a hot wire rake or grid, as shown in Fig. 10.14 and this depends on the type of flow to be measured.

Chapter 11
Non-intrusive Measurement Techniques

11.1 Basic Principle of Non-intrusive Techniques

Non-intrusive measurement techniques are based on optical techniques and have tremendous advantages, but as all measurement techniques they have their limitations and drawbacks. Aside the arduous technical issues associated with data processing and analysis, their main physical limitation is simply based on visual access of the experimental domain of interest and also the ability to guide the light/laser source for sufficient illumination. This depends a lot on the experimental facility and sometimes the working section has to be redesigned and equipped with sufficiently large windows with high optical quality glass or Perspex which is a strict requirement for classical interferometry techniques. For applications involving the use of invisible lighting, infra-red for instance, the windows have to be made of special material such quartz and zinc selenide (revisited in Sect. 9.5.3). This issue can be addressed by placing the instrumentations within the working section itself, if possible or it is easier done in experiment in open atmosphere.

The focal length of lenses needed for some optical techniques could restrict their use in some facilities with large working sections. Also, in high speed testing facilities mechanical vibrations and high intensity sound could damage the fragile optical components which tend to be quite expensive. During optical diagnosis, even if this sounds trivial, a high-quality window is critical to compensate for losses in illumination due to the development of wall boundary layer over the window, tracer particles seeded in the flow but more importantly optical aberration from the window itself.

Non-intrusive methods for investigation could be classified in three categories:

- techniques based on the change in the optical property of the fluid such as the refractive index due to a density or temperature of the flow, see Sect. 11.2;
- techniques based on tracking particles, either those present naturally in the fluid or seeded artificially, assuming that these particles are being convected at the same speed of the flow, these techniques will be presented in Sects. 11.3 and 11.4;

© Springer Nature Switzerland AG 2020   237
B. Chanetz et al., *Experimental Aerodynamics*,
Springer Tracts in Mechanical Engineering,
https://doi.org/10.1007/978-3-030-35562-3_11

– spectroscopic techniques based on the physical process during the intermolecular interaction between the illumination source and the fluid or the material used for the surface of the model; this will be covered in Chap. 12.

11.2   Interferometry

11.2.1   Light Interference and Refractive Index

The first category of techniques listed above is based on the deviation of the path of light due to the change in refractive index in certain region of a flow where the velocity changes rapidly. In a gas or air as in most aerodynamics applications the refractive index, n, and the density, ρ, are related by Gladstone and Dale's relation given as:

$$n = B\rho + A$$

where A and B are constants. Therefore, the variation in the optical property of the flow could be detected using a technique capable of sensing the change in density or vice versa. Based on this relation the region with variation in density, for instance in the presence of a shock or expansion wave, mixing zones and boundary layers, can be visualised and captured. This technique is an extension to the visualisation technique presented in Sect. 7.5.

Interferometry has historically been used to visualise compressible flows where spectacular images of the density field can be captured (see Sect. 7.5.3) and provide quantitative information which is very useful to study transonic and supersonic flows. The basic principle of interferometry is to capture the interference of two waves emitted from the same monochromatic, coherent source on a sensitive screen or nowadays straight onto a CCD sensor:

– the wave passing the working section, through the flow in concern,
– the wave passing outside the working section, through the unperturbed flow.

Both waves interacts on the optical sensor to produce an interference fringes pattern which are lines of equal phase along the optical path length, nl, the product of the refractive index, n, and the geometrical length of light beam, l. If the flow is uniform, with the interferometry system set-up properly, the optical sensor will be covered by a white fringe which corresponds to the constant phase. But once non-uniformity is introduced in the flow the density variation will result in a change in refractive and hence a deviation in the optical path and phase shift of the light wave. The shifted and unshifted beams interact to add or subtract energy to each other and this produces interference fringe patterns or the interferogram of the flow.

For a two-dimensional planar flow, in a working section with a width, b, the change in the length of the optical path introduced by the flow at all points is given as:

$$\Delta L = b\Delta n$$

By choosing a reference fringe, the change in the optical path between two points, R, and M, in the flow field can be expressed as a function of the fringe spacing:

$$\Delta L = \lambda(N_M - N_R) = b(n_M - n_R)$$

where λ is the wavelength of the light source, and N_R and N_M the fringe numbers between points R and M respectively. If the flow is two-dimensional, the density, ρ does not vary along the spanwise direction. Assuming that, $b = l$, the value of ρ at a point N_M within the fringe is expressed by the following relation:

$$\rho_M = \rho_R + \frac{\lambda}{bB}(N_M - N_R)$$

where ρ_R is the value of ρ corresponding to the reference fringe N_R. This value is determined by considering a fringe passing in a region of uniform flow where the temperature and pressure can be measured.

In practice each fringe, black or white, is identified by its median line where the light intensity changes from maximum to minimum or the two extremes of the grey scale intensity (in theory it should be possible to divide the main fringe into multiple fractions of fringes but this is not very accurate). The main issue with the interferogram is to identify the centre of each fringe; however this could be resolved more accurately nowadays using more advanced image processing routine. Once the centre of each fringe has been identified, the density field, $\rho(x, y)$ over a 2D plane can be determined within the experimental domain.

In the presence of a uniform flow where the screen is lit uniformly, this mode is referred as the infinite fringe interferogram as shown in Fig. 11.1a. In this set-up the fringes (or more precisely their centre) could be identified as lines of constant density, lines of iso-Mach numbers or iso-bars, if the flow is isentropic. If the variation in density is modest the number of fringes and their intensity is low and this reduces the

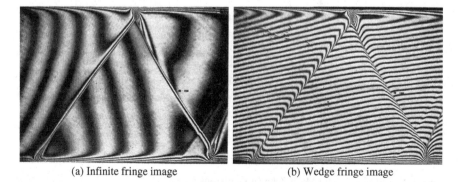

(a) Infinite fringe image   (b) Wedge fringe image

Fig. 11.1 Interferogram due to a reflected shock (© ONERA)

accuracy, for instance in transonic flows where ρ changes by about 10%. In order to increase the number of fringes, wedge fringe interferometry mode can be employed as shown in Fig. 11.1b.

Instead of both light waves being parallel as in the infinite fringe mode, one is tilted by a mirror and this introduces a linear variation in the optical path length while it passes through the flow. In the absence of a flow the black and white fringes are parallel to each other. The development of the flow introduces a change in the density which distorts the fringe pattern. This is captured on another sensing optic which is compared with the pattern obtained from finite fringe mode and used for calculation of density and further analysis.

Interferometry is employed for axisymmetric flows without major changes to the mode of operation, the main modification is to the image treatment as the relation between the fringe position and optical path length is different. Since the density of the flow field is not uniform along the span, it is calculated using the classical Abel's transform.

11.2.2 Mach-Zehnder Interferometry

The classical interferometry set-up follows that of the Mach-Zehnder type as shown in Fig. 11.2. Due to the very short coherence length of an ordinary white light the set-up is quite delicate. Therefore, the optical path lengths of the reference or reflected beam and the test or transmitted beam have to be within a few wavelengths with respect to

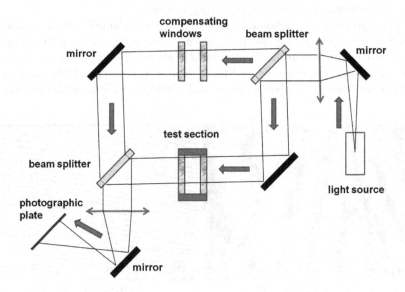

Fig. 11.2 Optical set-up for Mach-Zehnder interferometry

each other. Thus very high optical quality glass is required for the windows of the working section and usually an additional pair is required for the reference beam to account for losses due aberration from the windows.

11.2.3 Holographic Interferometry

Due to the larger distance of coherence of the light wave generated from laser sources, interferometry technique has been greatly simplified and this has attracted holographic techniques for creating interference patterns which can be treated using the same analysis process during classical interferometry. In the holographic technique the reference and the transmitted beam is created using either a helium-neon or argon laser. As shown in Fig. 11.3 the light beam is expanded through a pin-hole then it is collected by a spherical lens to have parallel light fringes.

The unexpanded reference beam which passes outside the working section through the unperturbed field. Outside the working section the transmitted scattered light is collected by another lens and is projected on a high resolution photographic plate, where it interacts with the reference beam also projected here and this creates a hologram. The interference pattern is then revealed by developing the photographic plate. To obtain an interferogram this process is repeated twice without removing the photographic plate, once without the flow and then with the flow field. Therefore, two waves are registered on the photographic plate one with the uniform field and the other with the perturbed field. By shining a laser light on the photographic plate both waves are reconstructed simultaneously leading to interference between the two holograms. The resulting image is captured to obtain the interferogram which is processed in the usual manner.

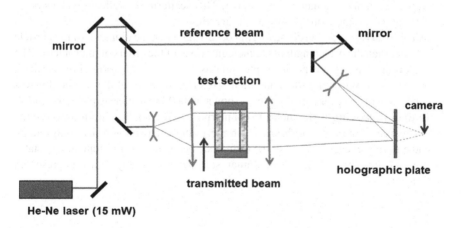

Fig. 11.3 Optical set-up for holographic interferometry

Holographic interferometry is easier to set-up than the classical interferometry, due to the extended distance over which a laser source preserves its coherence. Thus, the optical path length of the reference and the transmitted beam does not have to be adjusted very accurately as opposed to a white light source. Besides, a good quality glass will be sufficient for the test section windows and the lens and other instrumentations are more affordable.

Interferometry is a powerful tool for studying very complex compressible flow fields which are easily perturbed by solid probes. It's an accurate technique which can provide a considerable amount of information over a large experimental domain or field of view. However, its extent of application is reduced in the case of separated flow where the density is almost constant and it is impossible to make any measurements of turbulence. In practice, interferometry is limited to two-dimensional planar or axisymmetric flow. The extension to three-dimensional flow is possible through tomographic imaging, but the set-up, operational mode and process of image reconstruction, is so challenging that the technique has been applied only to very few rare cases.

11.3   Mechanism of Light Scattering

Light scattering is the property of particles that allows the reflection of a light wave in all directions. The scattering takes place without the loss of energy therefore the wavelength remains constant. The type of scattering depends on the form of the particles and their size, to distinguish between:

- **Rayleigh scattering** is the elastic scattering which occurs when the diameter of an assumed spherical particle is smaller than the wavelength of the incident wave, typically a tenth of a nanometre smaller. This scattering or reflection takes place without losses and isotropically in all directions.
- **Mie scattering** is the elastic scattering occurring when the diameter of the particle is larger than the wavelength of the incident wave, of the orders of micrometre. The reflection of the light is not isotropic and this depends on the form of the particles being spherical or cylindrical and their size. The intensity of the scattered waves depends a lot on the angle of scatter as shown in the Mie-scattering plot in Fig. 11.4.
- **Raman scattering** is the inelastic counterpart of Rayleigh scattering, where the energy is not constant. The difference in energy between a photon absorbing energy and emitting energy is equal to the difference between two different energy states of scattering object. The use of Raman scattering in aerodynamics is presented Chap. 12.

Fig. 11.4 Diagram of
Mie-scattering radiation
(incident light comes from
the left). *Image* Wikipedia

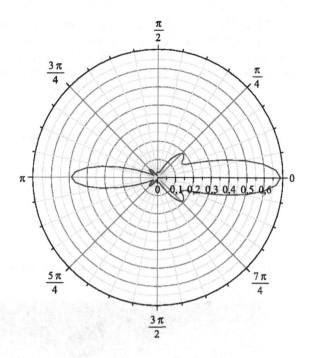

11.4 Laser Doppler Velocimetry or Anemometry

11.4.1 Basic Principles

The introduction of laser diagnostics systems in the 70s has allowed for the measurements and analysis of flow fields which was challenging using traditional techniques or sometimes impossible. Using Laser Doppler Velocimetry (LDV) or Anemometry (LDA), it is more reliable to probe separated flows in order to determine both the mean and the fluctuating velocity components. The advent of LDV almost coincides with entry in service of high-performance computers and thus the development of higher order numerical flow simulation tools, the birth of CFD. Therefore, progress in the flow simulation was a result of robust, parallel model validation through LDV measurements.

The basis of LDV is to measure the speed of micro-particles transported by a flow and this speed is supposed to coincide to that of the flow. However, this is not true for all types of flows, in rapidly accelerating or decelerating flows the particles cannot adjust their speed instantaneously to the flow speed and the particles have to travel a certain distance to so that they could catch-up with the flow. Such a scenario occurs downstream of a shock-wave, for instance, in the diverging region of a supersonic nozzle or in flow fluctuation at high frequency. While excluding the later cases, the measurements from LDV are accurate and reliable, regardless of the complexity of the flow phenomena being studied.

In laser Doppler technique a laser beam is split into two beams and using a focussing lens they are set to cross at a point to create a measurement volume within the region of interest in the flow. This is shown schematically in Fig. 11.5. Due to the high coherence of a laser beam, a fringe pattern is created at the crossing of the two beams within the measurement or probe volume. As shown in Fig. 11.6, while

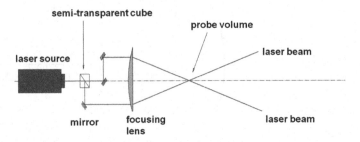

Fig. 11.5 Laser beams crossing forming the interference fringes of the emission part

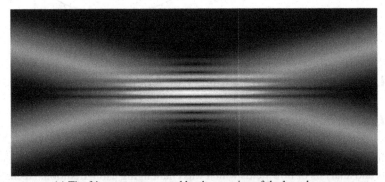

(a) The fringe pattern created by the crossing of the laser beams

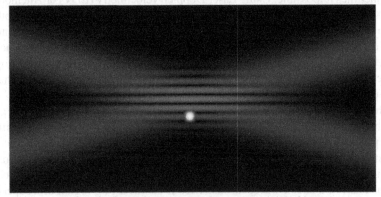

(b) The light scattered by a particle crossing a bright fringe

Fig. 11.6 The fringe pattern within the measurement volume (© ONERA)

a particle is crossing this fringe pattern, it is subjected to a dark or bright illumination, the intensity of the light scattered by the particle occurring at a frequency,
f_m, inversely proportional to the fringe spacing, i, and directly proportional to the
velocity component of the particle, U, which is normal to the fringe. As the fringe
spacing is defined by the optical characteristic of the laser source, f_m is captured by
a receiving optic and the velocity component can be obtained from:

$$U = i \, f_m$$

11.4.2   Signal Analysis

The light scattered by the particles is collected and transmitted to a photomultiplier
which converts it into an electronic signal. The passage of a particle within the probe
volume produces a signal or more commonly a "burst" as shown in Fig. 11.7. This
signal is constituted of a low frequency component (the pedestal) corresponding to the
Gaussian distribution of the light in the emitted laser beam, on which the periodic
signal containing the interesting information is superimposed, together with high

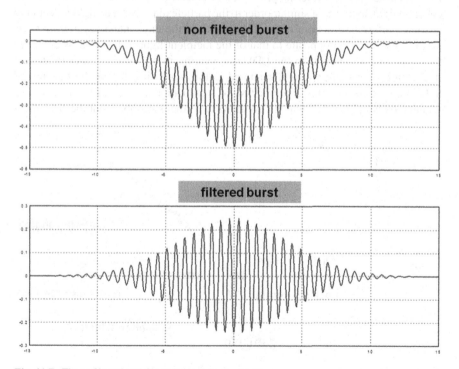

Fig. 11.7 The unfiltered and filtered signal from a burst

frequency noise. Just to have an idea, in a flow at 300 m/s and a fringe spacing of 15 μm, the frequency, *fm*, is occurring at 20 MHz and the duration of a burst is approximately 1 μs. Therefore, the issue at hand is to measure a high frequency, very accurately, in a signal of very short duration and potentially in the presence of noise.

Before further analysis, the signal is filtered by both a high-pass and low-pass filter to remove the low frequency, background noise, and the high-frequency noise respectively. In addition a minimum and maximum limit is set to eliminate high amplitude signals which are linked to large particles not convected by the flow and spurious signals. Various methods were developed to extract the frequency content of the signal and most of the systems are now based on Fourier transform of the signal.

11.4.3   Modes of Operation

The part that splits the main beam into two and the lens that focusses them into the probe volume is called the transmitting optics and the part that collects the scattered light, the receiving or collecting optics. The light scattering process that takes place in LDV is based on Mie scattering principle introduced above as the wavelength of the coherent light source is of similar order to the particle diameter. The light is scattered in some preferred directions defined by lobes and the most intense scattering takes place in the direction opposite to that of the incipient light. If this light is collected by the receiving optics then the LDV is operating in a forward scatter mode and if the slightly less intense light scattered in the same direction as the transmitted light is collected then this is the backward scatter mode. This has been illustrated schematically in Fig. 11.8, which also shows two other side lobes, symmetrical about

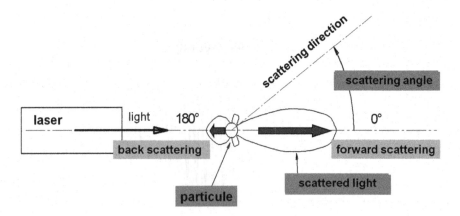

Fig. 11.8 Lobes of the light scattered by a particle

the axis of the probe volume and these are collected in the side scatter mode. The intensity of light in the forward scatter mode is about 100 times higher than the other modes.

- **In the forward scatter mode**, the receiving optics is positioned in the opposite side of the transmitting optics (incident light), where each element is placed on each sides of the working section. This set-up is more advantageous in terms of optical performance and it gives a higher signal to noise ratio and is recommended for measurement in high speed flows.
- **In the back scatter mode** the transmitting and receiving optics are on the same side of the wind tunnel and are usually mounted within the same compact mount which makes the alignment easier and hence ease of use. The main issue is the signal to noise ratio which deteriorates with flow speed and usually limits the use of this mode for speeds up to 300 m/s.

In order to determine the direction of the velocity component, the laser beams are passed through a Bragg cell (acousto-optics modulator) which in modern system acts as the beam splitter as well and applies a frequency shift to one of the beam. For an observer fixed in space, the oscillation in the fringes is equivalent to a shifted frequency corresponding to a velocity, $U_r = i \times f_r$. The frequency, f_r, is chosen in such a way that the relative speed, U_m, which is deduced from the frequency, fm, measured in the interference fringe pattern is always positive. Therefore the absolute velocity in the wind tunnel frame of reference is given by:

$$U_a = U_m + U_r = i(f_m + f_r)$$

Bragg cells are used even in the absence of reverse flows as it ensures that velocity components of small magnitude does not affect the accuracy of the measurement and also the particle crosses through sufficient number of fringes.

11.4.4 Multi-components Measurements

Two velocity components are measured by intersecting the pair of fringe patterns from two different lasers within the measurement volume and in this configuration a total of 4 beams are generated as shown in Fig. 11.9. Each pair of beams is different in colour and wavelength so that the collection optics can separate the two signals, a process essential for resolving the two velocity components. In more recent compact system both pairs could be transmitted from the same 'head' unlike in Fig. 11.9, including the collections optics used in back scatter mode. For the third velocity component, another pair of beam or fringe patterns is needed and this is usually transmitted from a separate head mounted off-axis from the two-component head. The set-up and alignment for 3-components measurement is more time consuming as all six beams have to intersect within the probe volume very precisely. Due to the increase in the measurement volume validation routine is employed to reject signals

Fig. 11.9 Two pairs of laser beams in an LDV set-up for two components measurements (© ONERA)

produced by passage of different particles. This ensures that all 3-components are related to the same particle and resolved simultaneously in time which is imperative for determining velocity fluctuations and the turbulence quantities.

11.4.5   Flow Seeding

The confidence level in the LDV measurement is usually set by the validation routine which relies on the quality and quantity of particles seeded or entrained in the flow. The choice of seeding particles is dictated by several factors:

- Size: Smaller particles tend to follow the path of the streamline more precisely however the intensity of light scattered is lower, hence weaker signal.
- Appropriately sized particles can bond to each other to form larger particles which affect the accuracy of measurement as they deviate from the actual streamline and are more problematic in rapidly compressing or expanding flows.
- For propulsion studies which involve heated jets, the particles need to withstand high temperatures without auto-igniting or vaporising.
- Some products used for seeding particles are not compatible with PSP and TSP
- Concerning, health and environmental impact, some particles can have harmful health and environmental effects. Therefore, one must ensure that the tunnel is

well sealed and the ejection is well controlled. Also, the personnel are equipped
with appropriate breathing equipment for prolonged exposure.
- Particles can also adhere to the windows of the test section and blur the view of
the collecting optics. This will result in a drop in sample rate, decrease validation
level and increase uncertainty.

Several types of particles were/are commonly used, a cheap one being incense
smoke. Others are atomised silicon, and olive oil and DEHS (di-ethyl-hexyl-sebacate)
which normally requires a high-pressure atomiser as well for oxides of aluminium
and zinc for high temperature flows. In addition, to the issue of particle entrainment
mentioned above, statistical bias can be introduced by different particle types depending on the location of measurements. In swirling flows or vortices, the centrifugal
force acts to eject the particles away from their centre resulting in poor seeding concentration in that region, an issue also encountered in separated flows due to the low
momentum of the fluid within the recirculation zone. These issues are encountered
during PIV measurements as well which will be covered in Sect. 11.6.

Figure 11.10 shows some early results obtained at ONERA using a two component
LDV system, downstream of a model which represents the rear of a body of revolution

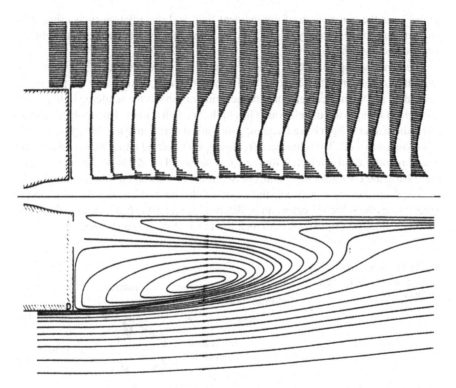

Fig. 11.10 The velocity of the flow field downstream of a body with a supersonic jet; the figure on
top shows the velocity vectors and the bottom the flow streamlines (© ONERA)

with a propulsive jet at the centre. The bottom figure shows the streamlines deduced from the measurement of the base flow.

11.5   Doppler Global Velocimetry

Planar Doppler velocimetry or Doppler Global Velocimetry (DGV) is a flow velocity measurement technique based on particle interference similar to LDV. However, unlike LDV which is a pointwise measurement technique, DGV can resolve the flow velocity over multiple points (of the order of hundreds of thousands) within a larger spatial domain. The basic principle is based on determining the Doppler frequency shift, Δf, in a light scattered by a particle moving at a speed $\vec{V}$. From Fig. 11.11, if $\vec{E}$ and $\vec{R}$ are the directions of the incident light and an observer respectively, the frequency shift is:

$$\Delta f = \frac{1}{\lambda_0} \vec{V} . \left(\vec{R} - \vec{E} \right)$$

where λ_0 is the wavelength of the incident light and this relation shows that Δf is proportional to the velocity of the particle, the direction of the incident light and direction of observation.

The basic set-up for DGV measurement is shown in Fig. 11.12, a light sheet illuminates the plane of interest in the flow and the light scattered by the particles is tracked by a reference camera and another camera records the scattered light passing through an iodine vapour cell which has high absorption lines. Due to the Doppler's effect the scattered light has a frequency shift and the transmission through the iodine cell changes as well. This change in frequency is converted into a change in intensity which is more easily detected by the CCD sensor. The Δf is determined by post processing the grey-scale intensities between the filtered and unfiltered image.

Based on the angle between the plane of the laser sheet and the direction of observation, the velocity vector can be determined and using 6 synchronised cameras all three velocity components, both mean and instantaneous can be determined in the

Fig. 11.11 Schematic representation of light scatter for DGV

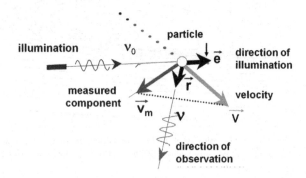

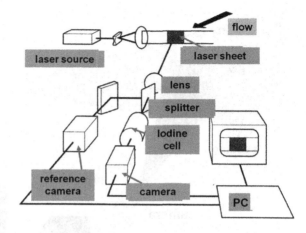

Fig. 11.12 Schematic representation of laser-optics setup for DGV

2D field. Alternatively, the laser sheet can be rotated about its axis or three laser sheets can be used, each at an angle, however limiting measurements to mean quantities only.

11.6 Particle Image Velocimetry

11.6.1 Basic Principle of Planar PIV

Particle Image Velocimetry (PIV) was implemented in the 90s and has since become one of the most common techniques used in fluid mechanics research. The basic principle is analogous to LDV where fluid flow is measured by sensing the displacement of artificially injected particles convected by the flow. In its most primitive set-up as shown in Fig. 11.13, an experiment using PIV consists of a laser sheet (about 1 mm thick) expanded by a cylindrical rod which illuminates the measurement plane. The laser is pulsed continuously at a very short time delay to light-up the particles seeded into the flow. A camera is positioned normal to the plane of the laser sheet and records the images of the particles displacement, at two time instants. The images are then processed to determine the displacement of the particles within the 2D flow field for the measurement of two velocity components in the plane of the laser sheet.

The quality of the measurements is dictated by the choice of particles which are transported passively by the flow. This behaviour can be quantified by Stokes number, which is the ratio of the characteristic time of particle (the time constant in the exponential decay of the particle velocity due to drag) to the characteristic time of the flow. For good quality and reliable PIV measurements the Stokes number of 0.05–0.1 is usually recommended. This pre-requisite on size and particle types has serious implications on the signal-to-noise ratio of the captured image. In water it is possible to use solid particles with a diameter of the order of 100 μm and a

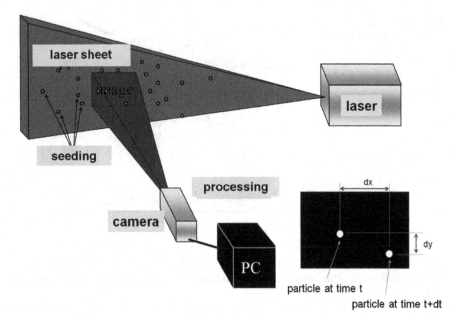

Fig. 11.13 Schematic representation of a planar PIV setup for the measurement of two velocity components

material density of the same order as water; these particles scatter a considerable amount of the laser light. It is within a regime of the Mie scattering process where the intensity of the light is proportional to the square of the diameter of the particle (usually larger than few μm) and this guarantees more pronounced particle images which are not masked by background noise. In air the particle seeding is through atomisation of natural or synthetic oils and the sizes are of the order of 1 μm; as mentioned previously oxides of metals are used for flow that undergoes very large amount of heat transfer.

11.6.2 Image Processing

Due to the miniature size of the particle, what is captured by the CCD sensor is not a magnified image of the particle but the bright spot (speckle) due to the scattered light, for a given aperture size on the lens. This is shown in Fig. 11.14. In air flow there is a compromise while adjusting the collection optics where a large aperture allows more light hence high intensity but reduces the size of the spot and vice versa. For an accurate tracking of the particle the spot needs to be spread over at least 2–4 pixels.

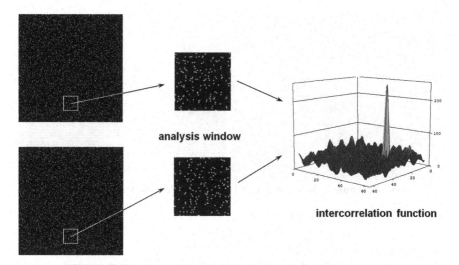

analysis window

intercorrelation function

Fig. 11.14 The steps for extracting the particle displacements from PIV images

The most commonly used algorithm to calculate the displacement of the particle is based on statistical correlation over a defined interrogation window which determines the spatial resolution of the measurement. Then the mean displacements within the neighbouring spots are determined by a cross-correlation of the spots within the interrogation window, for each image pair and this is repeated over the full sample of the image patterns captured. The displacements are identified as those with the peak in the correlation function, this process being summarised in Fig. 11.14 and the displacement is converted from pixelated to spatial domain using a calibration.

The pixel size of the correlation peak is approximately equal to the spot size of the particle. To determine the displacement with satisfactory accuracy a sub-pixel interpolation of the peak is recommended. This requires the base of the peak to be wide enough to avoid over-representation of the displacement; a bias referred as peak-locking. In order to reduce this bias effect, the aperture opening could be adjusted accordingly or the image could be slightly defocussed. Figure 11.15b shows the velocity field deduced from the raw image of the particles in the laser sheet shown in Fig. 11.15a.

In addition to many other applications, PIV is commonly used to study the structures within a complex, detached flow, for example in the wake of automobiles which contributes to more than 30% of the drag. Figure 11.16 shows the velocity contours along the symmetry plane downstream of a car model; the contour lines represent the mean velocity streamlines calculated from a large sample of images captured using PIV.

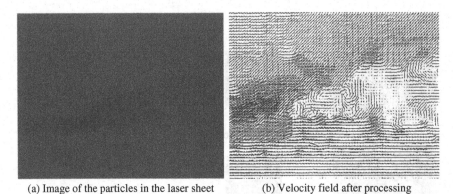

(a) Image of the particles in the laser sheet   (b) Velocity field after processing

Fig. 11.15 Result of PIV processing

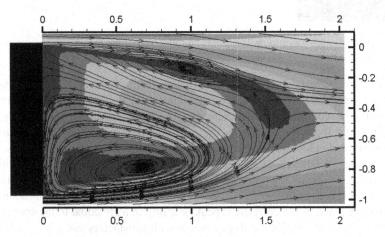

Fig. 11.16 PIV measurements in the wake downstream of a car model, showing the contours of the mean streamwise velocity (© PRISME Laboratory)

11.6.3 Three Components Stereo PIV

The advent of stereoscopic PIV in the mid-90s allowed for the measurement of three velocity components in the 2D plane of the light sheet. Here, the light sheet is slightly thickened with respect to the case of planar PIV so as not to lose particle pairs during the two pulses of illumination. The visualisation is through two cameras positioned each at an angle to the plane of the laser sheet, following a slightly more involved alignment process, a calibration and perspective correction the out-of-plane (normal to the plane of the laser sheet) displacement can be determined. A schematic representation of the stereo PIV set-up is shown in Fig. 11.17; this is for the study of the dynamics of fluctuating structures within a mixing layer generated from a round

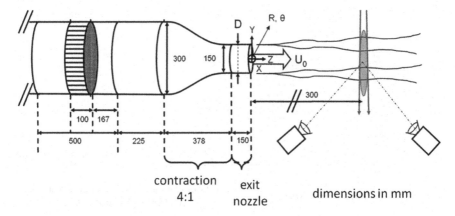

Fig. 11.17 Schematic representation of the setup for a time-resolved, stereo PIV measurement (© ONERA)

jet. PIV systems can be operated at high frequency of the order of 20 kHz. This allows for time-resolved measurements in low speed flows and for slightly faster flow the more energetic frequency band can be resolved for temporal analysis.

During the study presented in Fig. 11.17 the flow was seeded by DEHS particles, approximately 1 μm in diameter and was generated using a Laskin nozzle. The flow was illuminated by a Litron Nd-YLF laser with a wavelength of 527 nm, set to produce pulses at 2.5 kHz and at pulse delay of 60 μs between each image pair. The image of the flow domain of interest was captured using two Phantom V12.1 cameras. In order to maximise the light intensity, the cameras are set to operate on forward scatter mode with respect to the orientation of the light sheet. The cameras are equipped with Scheimpflug mounts to compensate for image sharpness losses due to perspective angle. Prior to measurements the cameras are calibrated by placing a calibration plate, made up of equally spaced dots, in the same plane as the laser sheet. This helps to convert from the distorted (diverging) image of the particles displacement seen by the cameras to the actual movement in the laser plane. While processing the images of the particles an auto-calibration process is employed to compensate for the uncertainty due to misalignment between the calibration plate and the laser sheet as mounting the calibration plate at the position of the laser sheet is challenging. Again using a statistical correlation algorithm the displacement field is obtained for a velocity vector every 5.4 mm, a value corresponding to the size of the interrogation window. The field of view covers approximately 1.6 times the diameter of the jet (see Fig. 11.18).

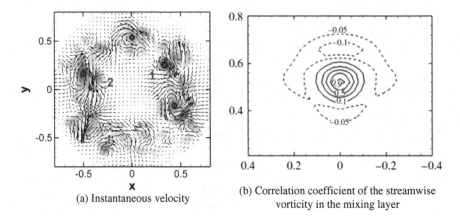

(a) Instantaneous velocity

(b) Correlation coefficient of the streamwise
vorticity in the mixing layer

Fig. 11.18 Measurement of the velocity fluctuation in the mixing layer of a turbulent round jet
using stereoscopic PIV for 3 components measurement (© ONERA)

11.6.4   PIV Tomography or PIV 3D

Around the mid-2000 tomographic or volumetric PIV started appearing; this enabled
measurement of a three-dimensional flow field (3D3C). As shown in Fig. 11.19, in
this technique the thickness of the laser sheet is increased using another set of lens to
produce a volume (10–20 mm wide) in which the flow is captured using 3–4 cameras
at least.

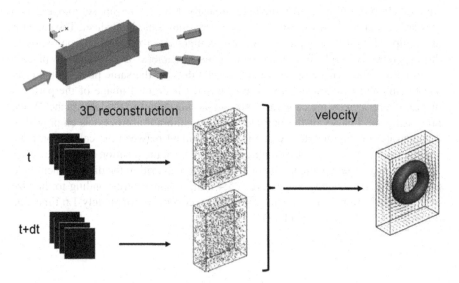

Fig. 11.19 Schematic representation of the processes involved in tomographic PIV

During one of its most common utilisation mode the results processing involves two main steps; firstly, a tomographic reconstruction which determines the intensity of the particles within a three-dimensional grid of voxels for all time instants. Secondly, the displacement is determined using an algorithm based on a three-dimensional correlation, using as input the volume of intensities as opposed to pixel intensities in planar PIV. This results in a three-dimensional velocity field of all 3 components.

This method has already been used previously to characterise canonical flows, but still, it is more delicate to employ as opposed to planar or even stereoscopic PIV. The set-up is bulkier and requires more space due to the additional numbers of camera, supports and traverse system for the calibration, for a relatively small measurement domain of $10 \times 10 \times 2$ cm^3. Besides, the accuracy of the volumetric measurements depends significantly on the seeding particle density and a high concentration can introduce a lot of noise, due to the presence of spurious peaks or ghost particles during the reconstruction of the volume intensity. A very low seeding density is detrimental either for correlation or to resolve the smallest length scale accurately in complex flows.

11.6.5 Particle Tracking Velocimetry

Particle tracking velocimetry (PTV) is based on following the displacement of individual particles in the flow within the interrogation window. Therefore, a particle tracking method is used which involves measuring the displacement of the same particle, based on the successive images which tracks its position at an instant in time (see Fig. 11.20). Here a relatively low seeding density is recommended, but it could be challenging to measure in dead air regions of detached flows, as the tracer particle have difficulty to penetrate this region. On the other hand, if the seeding density is too high the noise level might make it difficult to follow the tracer particle.

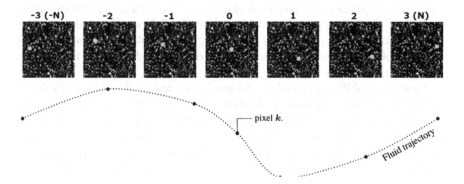

Fig. 11.20 Particle Tracking Velocimetry (PTV)

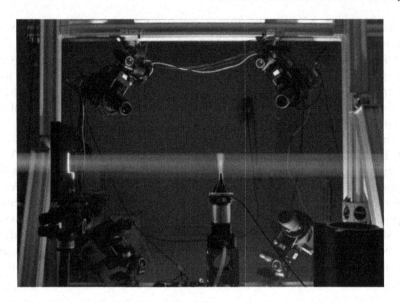

Fig. 11.21 Photograph of a 3D-PIV and 3D-PTV setup for the measurement in a low speed, subsonic jet (© ONERA)

Figure 11.21 shows the set-up for the measurement in a low speed jet using 3D-PIV and 3D-PTV, consisting of 4 cameras of 2048 × 2048 pixels each. The reconstructed volume contains 2239 × 2909 × 598 = 3.0 billion of voxels and Fig. 11.22 shows a comparison of the velocity field captured by both techniques.

11.6.6   Time Resolved Particle Image Velocimetry

The classical PIV described above captures the instantaneous flow field so that the successive images captured can provide information about uncorrelated phenomenon in the flow. This allows statistical analysis of the flow field where both the mean and RMS quantities can be obtained for determining the Reynolds stress tensors, but the evolution of the phenomena in time cannot be obtained and hence the frequency of their occurrence is unavailable.

Time-resolved PIV, which allows the study of the temporal evolution of a flow is a powerful technique for studying the flow mechanisms both in time and space and provides information about instability modes and turbulent structures. For a time-resolved PIV, a camera of high frame acquisition rate is required (100 k fps) and a more powerful laser pulsing at a speed of similar order. Again, the image pairs are correlated to extract the velocity and to obtain a time series of the flow field where

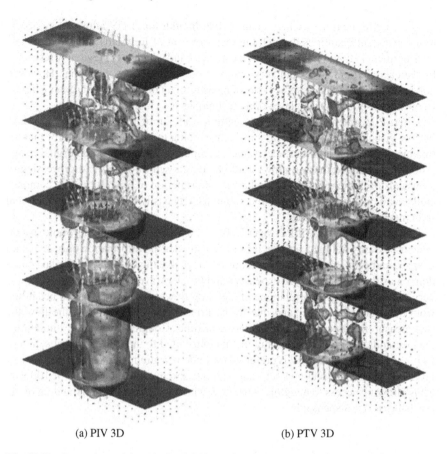

(a) PIV 3D   (b) PTV 3D

Fig. 11.22 Comparison of the velocity field in a subsonic jet captured using. **a** 3D-PIV and **b** 3D-PTV (© ONERA)

one can deduce the dynamic behaviour of the flow, in particular the turbulence in the frequency domain or spectra. The present practical limit of time resolved PIV is around 20 kHz.

11.6.7 PIV Vs. LDV: A Brief Conclusion

In its current versions laser-optic velocimetry started appearing in wind tunnel and fluid mechanics laboratories during the 80s. It has revolutionised our approach to conducting experiments by enabling velocity measurements in flow field without the use of intrusive solid probes. Regions of detached flows which were previously challenging to measure are now quantifiable. In addition, LDV allows more access to the turbulent quantities through the measurement of Reynolds stress tensors and measurement in flows within a higher frequency band which was previously restricted by data acquisition issues and very cumbersome treatment of the signal. However, hot wires still dominate in some types of measurement due to their unarguably higher frequency response but the laser-optic techniques are catching up steadily.

Later, PIV, where the basic idea is simple and can be traced back to early days of flow visualisation when it was subsided as being too labour intensive for manual treatment. Progress in optics, electronics, computational resource and digital image processing changed the way flows could be diagnosed through rapid processing of thousands of particle images in a relatively short period of time. It became commercially available very rapidly and thus found its way easily in many wind tunnels and fluid mechanics laboratories.

The major advantage of PIV over LDV is the ability to capture the whole instantaneous velocity field at a go, whereas LDV provides the information only at a point in space and a traversing process is required to resolve the flow within similar spatial domain as the PIV, leading to expensive and bulkier traversing hardware. Also, it is very difficult to derive correlations in unsteady flows as the probe has to be shifted from one point to another and due to the time-varying behaviour of the flow the two measurements are not representative in time. Coupled with these limitations, the cost and effort required for alignment contributed to the lack of prioritisation of LDV over PIV. But still LDV is preferred for finer measurements, such as in boundary layer traverse where it allows for near-wall measurements of the orders of 100 μm. Thus allowing access to a region of the flow where a "clinical" understanding of the mechanism is indispensable.

Chapter 12
Laser Spectroscopy and Electron Beam Excitation

12.1 Basic Principles

The advent of laser sources in the 1960s gave considerable impetus to the development of non-intrusive methods for the in situ determination of gas properties, including the velocity. These methods are based on physical processes related to the interaction between light and matter. The analysis of the resulting phenomena makes it possible to deduce the characteristics of the atoms and/or molecules composing the gas being tested and to measure properties such as its nature, concentration, energy levels, etc. Although these methods are not commonly used in aerodynamic applications because of their sophistication and limitations, they are powerful tools for studying complex flows. In particular, they give access to species concentration, local pressure, temperature and are of great interest for the study of very high temperature flows or flows containing chemically active combustion products.

Laser spectroscopic measurements are based on the interaction of a laser beam with some of the physical quantities in fluid. Depending on the interaction process, the laser light is either absorbed or dispersed by species active at the particular wavelength employed. The interactions between light and matter can be classified into three categories (see Fig. 12.1):

- In elastic scattering, light is diffused instantaneously without exchanging energy with the internal states of the molecules of the medium, the incident and scattered photons having the same energy. This is the case of Rayleigh scattering and Mie scattering.
- In inelastic scattering, the light exchanges energy with some of the molecules of the medium accordingly with the wavelength of the incident light. The molecule absorbs some of the energy and hence the scattered photon has less energy than the incident photon: this is the case of absorption, fluorescence or Raman Effect.
- In Raman scattering with anti-Stokes shift, the molecule loses energy and the scattered photon has more energy than the incident photon.

© Springer Nature Switzerland AG 2020
B. Chanetz et al., *Experimental Aerodynamics*,
Springer Tracts in Mechanical Engineering,
https://doi.org/10.1007/978-3-030-35562-3_12

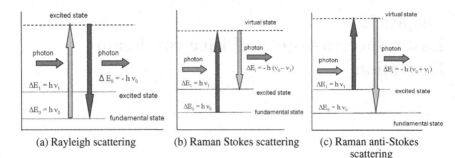

(a) Rayleigh scattering   (b) Raman Stokes scattering   (c) Raman anti-Stokes
                                                                        scattering

Fig. 12.1 Different light scattering mechanisms

The absorption and/or dispersion phenomena are characterised by an effective cross-section which measures the amount of radiation absorbed or dispersed in the process: the larger the cross-section, the higher the intensity, more radiation is absorbed or emitted. In both cases, the intensity of the resulting signal depends on the population density of the atomic or molecular energy states responsible for the interaction. In addition, the radiation resulting from the interaction is a function of the spectral properties of the interacting species and their molecular energy states. Thus, measurements are made on populations for which it is possible to deduce the temperature of the gas and the local concentration of the species, by analysing the spectrum of the emitted light. From these primary parameters, the other thermodynamic quantities, such as density and pressure, can be deduced using the appropriate laws of thermodynamics. The velocity of the gas can also be obtained by less direct methods.

Since the thermodynamic properties of a species are related to the spectral properties of the radiated signal, the determination of two quantities (for example temperature and pressure) requires the measurement of at least two spectral characteristics. In some cases, the variation of a single spectral characteristic is sufficient to determine the thermodynamic conditions of the flow. The intensity of the radiated signal gives a measure of the concentration of the species (or density in number of atoms or molecules). The temperature is most often deduced from the broadening of the spectral content by the Doppler Effect induced by excitation of atoms or molecules, the energy contained in this movement being proportional to the square root of the translation temperature. The velocity of the flow is determined from the shift in the centre frequency of the signal due to the Doppler Effect produced by the overall motion of the gas. The rotation and vibration temperatures can also be determined from a signal analysis. Figure 12.2 summarises the principle of the method.

Laser spectroscopy techniques have several advantages. They are non-intrusive and well adapted to the measurement of thermodynamic parameters. Most of the selected species have indeed a good spatial and temporal resolution allowing characterisation of three-dimensional flows. They do not require seeding of the flow by

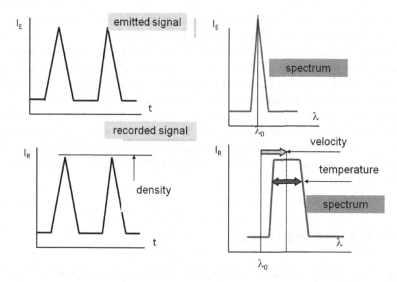

Fig. 12.2 Flow analysis through spectroscopic analysis

particles which may not be convected at the same speed. On the other hand, they require optical access and are generally complex to implement, their use being most often limited for fundamental research facilities.

12.2 Laser Absorption Spectroscopy

In this technique, the laser beam is tuned to a wavelength in resonance with an absorption range of a selected species. The attenuation of the beam passing through the test area is measured, the absorption of two or more wavelengths are used to determine both the temperature of the gas and the density of the absorbing species. This technique is simple to implement because it only requires optical access just enough to allow the laser beam to penetrate the measurement area. The generation of a detectable and usable signal does not depend on the power of the laser therefore low power lasers can be used. As a result, optical fibres can be used to perform measurements in internal flows or in areas of poor optical access. In principle, the laser absorption technique allows the measurement of the species concentration, the temperature, the pressure and the velocity of the gas.

In contrast, the method lacks spatial resolution, a measurement representing an integrated average along the path of the beam through the measurement zone. This limits its use to uniform two-dimensional flows along a transverse direction, or to axisymmetric flows whose local properties can be extracted by an Abel inversion. For a three-dimensional flow, this disadvantage can be overcome but at a cost due to complication of the technique, therefore favouring other techniques. In addition, the absorbing species must be present with sufficient density to have enough absorption for accurate measurements at high signal to noise ratio.

Among the laser absorption methods, the laser diode absorption technique is a powerful tool for the time-resolved measurement of temperature, velocity and concentration in a flow where requirements in terms of spatial resolution are not strict and where the line of integration inherent to this technique is not a problem. The technique consists of illuminating a cross section of the gas to be studied by an infrared light beam. Doppler broadening and shift in wavelength of the species absorption lines are used to determine the translation temperature and the flow velocity. With proper calibration, the area beneath the absorption lines gives a measurement of species concentration.

12.3   Rayleigh Scattering

Rayleigh scattering is through elastic radiative interaction (scattering) with the gas where the light from a laser beam is scattered at approximately the same frequency as that of the incident light (see Fig. 12.3). The intensity of the scattered light is proportional to the sum of the densities of all the species in the gas, weighted by their corresponding Rayleigh scattering cross-section. Spatial resolution is enhanced by observing the light scattered by a small region of the beam or by imaging the scattering of a two-dimensional light sheet. Density, pressure, and gas velocity can be determined by measuring the intensity of scattered light and its spectral properties. Rayleigh scattering is probably the easiest method to perform a local measurement of flow properties since it does not rely on spectral resonances of a seeding material (see LIF below). All species of a gas disperse the laser light and the wavelength of the laser is (in principle) irrelevant.

One of the drawbacks is that there is no spectral difference between the light diffused by the gas and the background light diffused by the other objects that can illuminate the measurement volume and the material constituting the optics. In addition, the flow must be free of solid particles whose cross section would be much larger than that of the atoms or molecules. It is thus difficult to distinguish the useful signal from the stray light which may be of equal or greater intensity. The intensity of

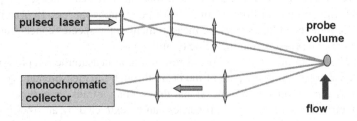

Fig. 12.3 Bench setup for Rayleigh scattering

the scattered light being inversely proportional to the fourth power of the wavelength of the laser, it is advantageous to work with a short wavelength, particularly with ultraviolet lasers.

12.4 Raman Scattering

Raman scattering is an interaction in which light from a laser beam is elastically dispersed by the gas at wavelengths shifted from that of the incident light. The cross section of the Raman scattering is much smaller than that of the Rayleigh scattering, making observation of the radiated signal more difficult. With this technique, it is possible to determine the species concentration, the temperatures of vibration and rotation of the particular species and the gas velocity.

The principle of the method is as follows (see Fig. 12.1b). When a photon hits a molecule, it leaves a fraction of its energy to the molecule that is raised to a higher energy level. When de-excitation occurs, a photon is emitted at lower energy than that of the incident photon (longer wavelength). This process is called Stokes-Raman scattering. The scattered light has a spectrum whose frequencies are characteristic of the molecule. In particular, the distribution of the intensities between the frequencies of the radiated signal depends on the initial distribution of the energy states in the gas under test. Thus, the analysis of the emitted spectrum constitutes a means of measuring the vibration and rotation temperatures in this particular state. The concentration of the species is determined from the amount of light contained in a certain narrow spectral band.

If the photon emerges from the interaction at a shorter wavelength, hence with higher energy, the process is called anti-Stokes-Raman scattering (see Fig. 12.1c). This diffusion process takes place with molecules of a higher energy level whose relative number is small, the anti-Stokes Raman signal being thus weaker than the Stokes signal. The use of stimulated Raman scattering mitigates this disadvantage (see Sect. 12.5 below). The technique of basic Raman scattering, or spontaneous Raman scattering, can be implemented with a laser having an arbitrary wavelength. However, it is advantageous to use short wavelengths, as for Rayleigh scattering. Pulsed lasers are often used because they can provide high power in a small measurement volume. The Raman signals, being at a wavelength different from that of the laser, are very little affected by a scattering background. In addition, the Raman effect is an interaction due to the radiation field and, for this reason, is not affected by quenching (a non-radiative energy exchange by collision between atoms or molecules which tends to decrease the radiated energy, see Sect. 9.3). The main disadvantage of this technique is the weakness of the signal. In addition, the laser beam being dispersed in all directions in space, the spatial resolution is obtained by observing the signal from a small region of the beam. Thus, the efficiency of spontaneous Raman scattering at visible wavelengths is generally very low (about 1000 times less than Rayleigh scattering). Stimulated Raman spectroscopy has been developed to eliminate this major drawback.

12.5   Stimulated Raman Scattering

This technique is analogous to spontaneous Raman scattering, the diffusion being produced by a first laser, the pump laser. The system includes a second laser, the probe laser or tuneable Stokes laser, with shifted frequency so that the difference in wavelength with the pump laser matches a resonance frequency of the molecule (see Fig. 12.4).

This arrangement is used in Coherent Anti-Stokes Raman Scattering (CARS) in which measurements are made with anti-Stokes radiation. In this process, the gas is illuminated by the pump laser at a frequency, f_1. The Raman scattering is stimulated by a tuneable probe laser whose frequency f_2 is adjusted so that the difference ($f_1 - f_2$) is equal to the frequency associated with a certain energy state of the molecule. The resonance interaction induces a strong anti-Stokes Raman dispersion at a frequency, f_3, such that $f_3 = f_1 + (f_1 - f_2)$. The CARS main advantage is that the scattering cross section is several orders of magnitude larger than the spontaneous Raman effect. In addition, the light emitted is in a preferred direction defined by the directions of the incident beams. As a result, useful light is collected more efficiently than in ordinary Raman scattering. The analysis of the CARS signal makes it possible to determine the nature of the species, its concentration, the temperature etc. The gas velocity can be measured from the Doppler shift. There are several variants of CARS, for example the double-line CARS (DLCARS) where four beams are used to excite two energy levels of the studied molecule which allows a more direct determination of the density and temperature of the gas.

Due to the intensity and high directivity of the transmitted signal, CARS is widely used for measurements in flames. Figure 12.5 shows the arrangement of the receiving part of a CARS bench used for hypersonic measurements. Figure 12.6 shows a comparison between density and temperature distributions in front of a cylinder placed in a Mach 10 flow, measured by DLCARS and computed by a Navier-Stokes code.

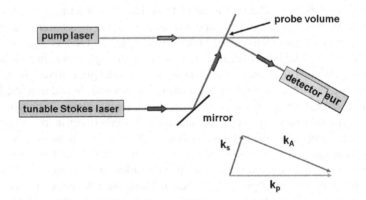

Fig. 12.4 Bench set-up for CARS analysis

Fig. 12.5 CARS bench installed in the R5Ch hypersonic wind tunnel at ONERA, Meudon; receiving part (© ONERA)

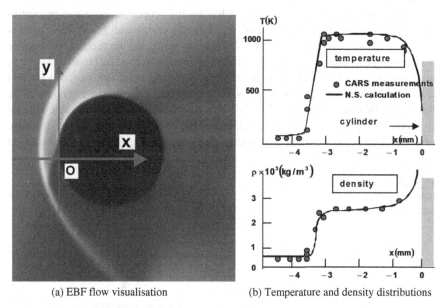

(a) EBF flow visualisation   (b) Temperature and density distributions

Fig. 12.6 DLCARS measurements in front of a cylinder at Mach 10 in the R5Ch wind tunnel at ONERA, Meudon (© ONERA)

12.6 Laser-Induced Fluorescence (LIF)

Laser Induced Fluorescence (LIF) belongs to the same category of radiation/absorption processes as employed in previous techniques; but here the measurement signal is obtained from the subsequent spontaneous emission of the absorbed energy or fluorescence (see Fig. 12.7). In this process, the emission takes place after a relatively long time, several seconds in some cases, whereas in other interactions the emission occurs after 10–8 s for most molecules. Because of this long relaxation time, the signal analysis must take into account the effect on certain species of non-radiative energy transfer by collision between the molecules (quenching). This phenomenon is in competition with the process of energy emission by relaxation, thus reducing the intensity of the fluorescence signal. Quenching depends on the temperature and concentration of the species.

To implement LIF, the laser beam, tuned to a resonance wavelength of the absorbing species has to excite a fraction of this species in the higher energy state of the radiative transition. The excited species then spontaneously radiates the energy absorbed (which is not lost through other relaxation pathways) at wavelengths allowed by the fluorescence spectrum of the excited species. Spatial resolution is obtained by observing fluorescence from a small region of the laser beam or by imaging the entire fluorescence from a laser plane. By analysing the fluorescence signal, the species concentration, the gas temperature and pressure, and the flow velocity can be determined.

LIF requires the presence of species having the proper fluorescence properties in terms of signal strength, spectroscopic characteristics and absorption capacity at wavelengths accessible by a laser. The species must also be thermodynamically coupled to the flow in a well-known manner. In flows with chemical reactions, a large number of the reaction products provide acceptable species for LIF, hence its interest in the characterisation of propellant jets. For aerodynamic applications in nonreactive fluids, the flow must be seeded with a low concentration of an adequate species (sodium, iodine, nitric oxide, acetone, etc.). Due to the toxic and corrosive nature of some of these species, aerodynamic studies with seeding should be performed in

Fig. 12.7 Fluorescence induced by laser or LIF; energy exchanges

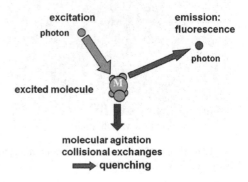

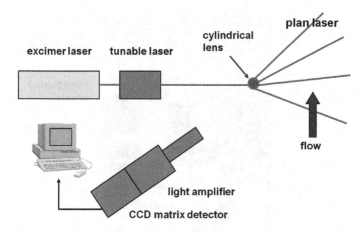

Fig. 12.8 Principle of Planar Induced Fluorescence, or PLIF

small facilities; except if acetone is used. The quenching phenomenon further limits the application of LIF to pressures below atmospheric pressure.

Fluorescence properties are also used to obtain an image of a flow region using the Planar Laser Induced Fluorescence or PLIF (see Fig. 12.8). In this technique, the area of interest is illuminated by a laser sheet and the fluorescence image recorded by a camera containing a two-dimensional matrix of photo-detectors. In general, the camera records and digitises the fluorescence intensity associated with the absorption for a single spectral function; hence the ability to determine the velocity, the temperature or the concentration fields according to the chosen function.

The flow velocity is deduced from the overall Doppler shift of the radiated signal. To accurately determine velocity, this method requires either measurement techniques sensitive to small frequency variations or sufficiently large flow velocities. To allow measurement of low velocities, a flow marking technique can be used in which a small volume of fluid is marked by a laser-induced process. Due to its long relaxation period, LIF allows tracking this volume convected downstream after a specified time interval. With this technique, velocities can be measured without strong requirements on the spectroscopic qualities of the signal. In this application, LIF is superior to LDV because it measures the velocity of atoms and molecules whereas LDV gives the velocity of particles much larger than the constituents of the gas. In practice, the good functioning of LIF remains delicate and complex. In addition, LIF only operates in a well-defined regime of pressure.

12.7 Electron Beam-Induced Fluorescence (EBF)

Electron Beam Fluorescence (EBF) is a technique well-suited for non-intrusive local measurements of density, vibrational and rotational temperatures in a low-density

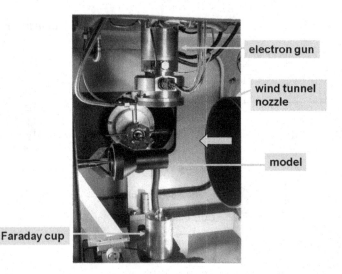

Fig. 12.9 Electron gun installed in a hypersonic wind tunnel (© ONERA)

nitrogen or air flow (up to 1015 molecules/cm^3). An electron gun creates an energetic beam of electrons (typically 25 keV) inducing a complex set of excitations in the gas along its path, some of these excitations producing radiative emissions over a wide spectrum from visible to ultraviolet light. Figure 12.9 shows an electron gun installed in the test section of a hypersonic wind tunnel.

Tomographic imaging by a scanning electron beam is a classic application of EBF (see Sect. 7.6.2). For this application, the electron beam is deflected by means of electrostatic plates at a repetition rate of 50 Hz to create a viewing plane. The technique is based on the formation of N_2^+ ions excited by the electron beam that traverses the flow. The almost immediate fall to a lower energy state gives rise to a fluorescence whose intensity is proportional to the gas density. At high densities, quenching destroys the linearity of the response; however the process can still provide qualitative information. Exposure times of several seconds are required for photographic recording. Figure 12.10 shows an EBF visualisation of a Mach 10 flow around a spatial probe model.

12.8   Electron Beam Induced Glow Discharge Measurements

This technique uses a miniature pseudo-spark type electron gun with the objective of measuring the boundary-layer velocity profile. The miniature pseudo-spark developed at ONERA generates an intense pulsed electron beam emitted by an electron gun. The beam enters the flow from a 0.3 mm hole drilled on the model surface and

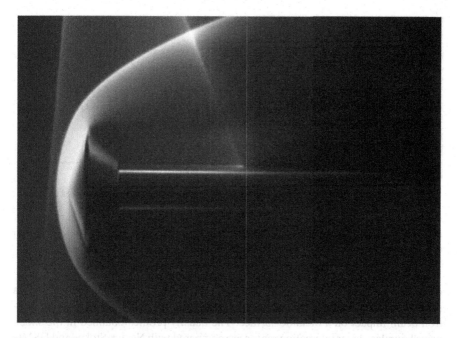

Fig. 12.10 Electron beam visualisation of the flow past a spatial probe model (© ONERA)

follows the path of a high-voltage glow discharge for 10 ns. The discharge filament is instantly connected to a high voltage capacitor via a thin metal rod parallel to the flow located at 100 mm from the electron gun outlet. This allows maintaining a very bright gaseous filament for a few microseconds, the initial rectilinear path of the discharge following closely the flow streamlines. With a precise delay time (5 μs) after activation of the electron gun, a CCD camera is briefly opened (250 ns) to record the position of the light column convected by the flow. The local velocity of the flow as a function of the distance to the wall is deduced from the horizontal displacement of a given point during the selected delay time.

12.9   Detection of X-ray Emission by Electron Beam Excitation

To perform density measurements even in case of quenching, a variant of EBF is used based on the detection of the bremsstrahlung effect and of characteristic X-rays. This radiation is emitted by electrons that are decelerated when passing close to an atom. The method has the advantage that the signal is emitted instantaneously and does not exhibit collision quenching. X-ray radiation at the measurement point is collimated and detected by X-ray sensors equipped with preamplifiers. The photograph in Fig. 12.11a shows the electron beam used to perform X-ray measurements

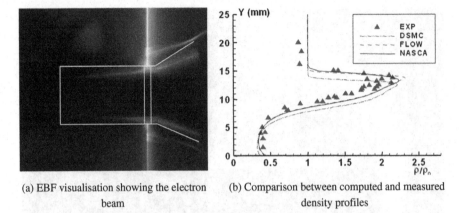

(a) EBF visualisation showing the electron   (b) Comparison between computed and measured
beam   density profiles

Fig. 12.11 Measurements by detection of X-ray radiation in a Mach 10 flow in the R5Ch wind tunnel at ONERA, Meudon (© ONERA)

on a flared-cylinder configuration. The beam passes through the model (through a small tube) to avoid the intense production of X-rays that would result from the electron beam impact on a metallic surface. The density profiles obtained in the shock wave boundary layer interaction region are compared with Navier-Stokes and DSMC calculations in Fig. 12.11b.

Chapter 13
Computer-Aided Wind Tunnel Test and Analysis

13.1 Experimental Versus Numerical Analysis

Currently in aeronautics the aims and objectives of wind tunnels experiment are diverse. It could be purely for optimising the aerodynamic design at the high speeds (transonic for the transport aircraft) by conceiving efficient wing sections, coupling the structure and aerodynamics (optimised volume, lighter structure, aeroelastic behaviour), developing innovative deployable surfaces to achieve high lift at low speed, controlling the plane's behaviour (performance, manoeuvrability and flight envelope) up to maximum lift coefficient and beyond stall. In the field of terrestrial vehicles, the objectives are to decrease the drag in order to reduce the fuel consumption, to guarantee the stability of the vehicle (response to cross wind) and to increase interior comfort by reducing aerodynamic noise.

Figure 13.1 synthesises the contributions of flight and wind tunnel tests, and numerical simulation (CFD), based on their representativeness of the physical phenomena, the determination of aerodynamic forces, the analysis of the flow and the time cycle of design which could range from several years for the development of a new prototype to a few minutes for basic CFD computations.

Up to the 70s, theoretical and experimental (Experimental Fluid Dynamics or EFD) methods were mainly employed to predict the aerodynamic characteristics of a vehicle. Since then, Computational Fluid Dynamics (CFD) has gained in importance in the aerodynamic prediction due to significant progress in the field of numerical techniques and computer processors speed. At present, one can think that the contribution of CFD to the aerodynamic design is comparable to that of experiment, so there is a tendency towards an integration of EFD with CFD to improve design methods.

Wind tunnel testing has evolved significantly, in particular following the opportunities given by CFD and progress in testing and measurements techniques. So, it is now essential to establish a strategy for combining CFD and tests in the design phase of a new aircraft or ground vehicle.

© Springer Nature Switzerland AG 2020
B. Chanetz et al., *Experimental Aerodynamics*,
Springer Tracts in Mechanical Engineering,
https://doi.org/10.1007/978-3-030-35562-3_13

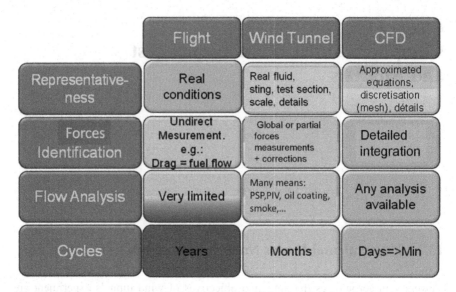

Fig. 13.1 The processes involved in the design of a vehicle (aircraft and cars)

In the experimental approach, the critical decision is to select the sequence of tests most adapted to the available wind tunnels, whereby the results of the wind tunnel tests having to consolidate the credibility of the program for the shareholders. The choice of adequate wind tunnels testing in the phase of design is essential and the adhesion to a strategy for the use of the wind tunnels must be a managerial decision with the value and the cost as the essential arguments. Improvement of effectiveness, accuracy and reliability of the tests is looked for by coupling EFD and CFD. Moreover, this strategy can be used for assessing the prediction of the aerodynamic characteristics in real flight conditions, on the basis of both ground tests and CFD.

As seen in Chap. 2, the wind tunnel tests differ from real flight in infinite atmosphere because of (see Fig. 13.2 for further information):

- confinement in the test section, effects of the model support and differences in Reynolds numbers,
- the lack of representation of geometrical small scale details,
- the fluctuations of flows induced by the motorised fan,
- deformation of the model subjected to the aerodynamic loads.

On the other hand, the limitations in CFD are as follow:

- issues in the representativeness of the computation results, because of a simplified turbulence modelling, boundary layer transition, separation, chemical reactions in hypersonic flows,
- time required for the generation of suitable meshes,
- relatively long computational time for a precise and reliable prediction.

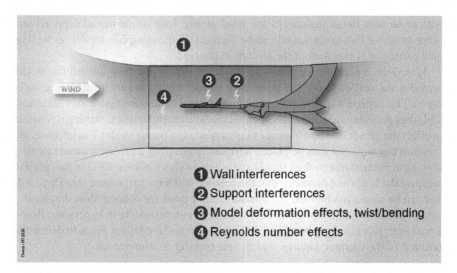

Fig. 13.2 Wind tunnel conditions differ from flight conditions

Progress in CFD in terms of improvement in accuracy and extension of the fields of application has been due to rigorous comparisons with experimental results. These comparisons are in general loosely coupled, with calculation on one side and experiment on the other, without a reciprocal process. However, it is frequent that comparisons are affected by light variations of the flow and ambient conditions, the geometry of the model, by slight variations in wind tunnel operating conditions, the deflection of model support, or its deformation under the aerodynamic forces to which it is subjected. In some cases, analysis of the experimental results cannot take into account the aerodynamic interferences due to the walls of the test section and/or the model support. Whereas, the computational mesh does not include test section walls and the support to reduce computational cost and time. It is even more difficult to compare the position of laminar to turbulent transition obtained experimentally against CFD. Such incompatibilities between experiment and simulation make it difficult to identify remaining problems for improving prediction methods.

13.2 CFD for the Preparation of Wind Tunnel Tests

The integration of a model in a wind tunnel test section requires taking into account geometrical constraints (wind tunnel size, influences of the model support, etc.) but also physical factors (impact of the model shape on the aerodynamics of surrounding surfaces, noise, etc.). Thus, CFD is used to define the model shape in order to consider these aspects before manufacturing the model. This helps in identifying the potential constraints due to the model installation, ensures and validates the representativeness of the model compared to the real flight conditions. These simulations can be realised with or without detail features of the wind tunnel, such as the model supports and other intrusive instruments, and can be focused on the complete model, as well as on a particular component of the aircraft in concern.

Moreover, a preliminary access to the behaviour of the flow around the model provides the test engineer the possibility of identifying the interesting or critical zones. This is particularly true when unconventional or new configurations are tested. CFD thus provides an initial assessment of the flow characteristics, making it possible to direct the choice and positioning of instrumentation on the model. For instance, the pressure tappings can thus be placed adequately ensuring a precise measurement of very local phenomena such as rapid expansions or shock waves. CFD is also used to estimate the loads and moments that the balance will have to measure (see Fig. 13.3) and can be useful in choosing the measurement planes for detailed flow diagnostics using optical techniques such as PIV or LDV. For experiments in hypersonic flows, a multi-physics coupling between CFD and a thermal modelling helps to define the position of the thermocouples for local heat transfer measurements.

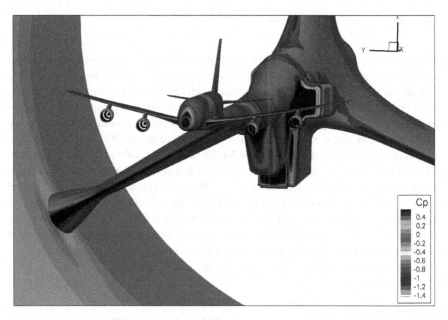

Fig. 13.3 CFD simulation of the model and its support in the ONERA S1MA wind tunnel (© ONERA)

13.3 Correction and Monitoring of Wind Tunnel Results by CFD

Initial calculation allows for correction of the effects resulting from wind tunnel walls and the model support which affect loads measurement and the pressure distribution on the model. The blockage effect can be accounted for and the Reynolds number, which is often much lower in the wind tunnel than in reality, can be adjusted accordingly. Figure 13.4 shows a scheme to extrapolate from the wind tunnel situation to the rigid model in free flight at the actual Reynolds number.

Instrumentation of models for wind tunnel tests is a complex task as there are major constraints on available space and due to extreme test conditions, such as ambient pressure and temperature, but not limited to vibration tending to perturb measurements. The availability of CFD as a complementary and independent source of information is then useful to validate the experimental results, even if the computed results are not perfectly reliable in absolute value. Even qualitative information can be enough to confirm or inform a trend observed during the tests.

A typical method of monitoring wind tunnel results is based on live pressure measurements. Since a model can be equipped with hundreds of pressure tappings, some of them might develop a fault during testing. The fault detection process could be a long and difficult task requiring a detailed examination of a considerable quantity of results. By simultaneously plotting the pressure distributions measured in the wind tunnel and comparison with CFD results, one can detect erroneous measurements in an easier and more reliable way. In spite of a large number of tappings, the spatial resolution in pressure measurements is still not fine enough to determine accurately the load distribution on certain parts of the model by integration. However, from CFD the pressure distribution can be determined with a much finer resolution. Even if the absolute pressure levels from simulations deviates slightly from those measured in the wind tunnel, the CFD results can be used to guide interpolation or extrapolation of wind tunnel measurements in poorly resolved regions due to lack of pressure tappings. To summarise, CFD allows the refinement of the local pressure distribution on the model, which is coarsely resolved by wind tunnel instrumentations.

Figure 13.5 shows a study of the influence of the model support on the wall Mach number distribution (deduced from the pressure on the wall by an isentropic relation)

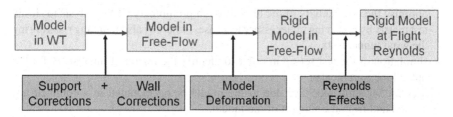

Fig. 13.4 Coupling tests and calculations for the correction of solid blockage due to model and support

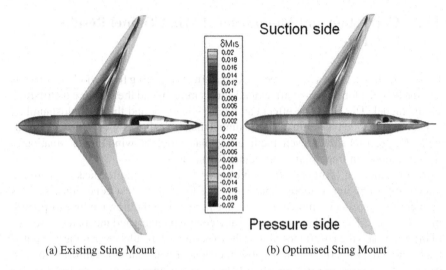

(a) Existing Sting Mount   (b) Optimised Sting Mount

Fig. 13.5 The Mach number difference due to the supports of the model in the European transonic wind tunnel ETW (© ONERA)

over an aircraft model tested in the European Transonic Wind Tunnel (ETW). Two forms of supports were considered: Existing Sting Mount and Optimised Fin Mount designed to minimise the distortions of the flow field on the wing and the fuselage tail section. Calculations allowed for the corrections on the upstream Mach number and the angle of incidence, hence accounting for the support effects on the aerodynamic loads.

13.4 Towards the Hybrid Wind Tunnel

The concept of hybrid wind tunnel addresses the above issues by tightly coupling experiment and numerical simulations with an aim of improving productivity, as well as the accuracy and reliability of the tests for vehicle design. The main aims are as follows:

– optimise the productivity of the tests through preliminary CFD analysis to shrink test matrix by removing unnecessary test cases, whereby also reducing risks during the wind tunnel tests,
– simulation of the test set-up in order to identify the region of interest for further optical measurements,
– account for solid blockage due to the model being mounted in a confined test section and from the support of the model,
– quasi real-time comparison of EFD and CFD results in order to consolidate the major findings,

- accelerate the processing of the large amount of data generated by methods such as PIV, PSP, TSP and measurements of model deformation,
- optimise the number of parameters for numerical simulations, such as the turbulence model and the computational mesh,
- consolidate the database constituting numerical simulations and test results obtained under identical conditions in order to improve the accuracy of CFD prediction.

To fulfil the above aims, the hybrid wind tunnel data acquisition and monitoring computer system must have a fast CFD solver associated with a robust tool for automatic mesh generation. Figure 13.6 shows the organisation of a hybrid system fulfilling these functions. Having defined the model's geometry, the "digital" wind tunnel on the right part of the figure carries out a pre-calculation including the model, the wind tunnel and the support. The computation results are transferred to the "analogue" wind tunnel, i.e. the real wind tunnel, in the left part of the figure. These results are used to optimise the model design and the test program. During wind tunnel tests, the results of the optical measurements are practically processed in real time and transmitted to the remote users as well as the wind tunnel team. These results, including measurements of model deformation, are sent to the "digital" wind tunnel for a new optimisation of the test parameters taking into account the model deformation. Finally, the results of the wind tunnel tests and revised calculations will be obtained for identical ambient flow and boundary conditions. The two sets of results are then combined to obtain the most probable aerodynamic characteristics through data analysis and assimilation (see below).

The digital wind tunnel makes it possible to evaluate the aerodynamic forces from individual elements of the full model by integration of the forces separately on

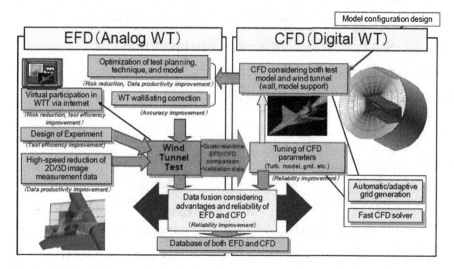

Fig. 13.6 Organisation of the hybrid, digital/analogical wind tunnel (© JAXA)

the fuselage, the wing, the nacelle, the pylon, the model support, etc. The user can thus examine the influence of the support on each component starting from pre-test calculations. Thus, before even manufacturing the model and the supports, the digital wind tunnel allows evaluation of the effects of the supports on the flow around the model. Figure 13.7 shows an example of the mesh for two types of sting mount and their interference on the global flow field can be compared in Fig. 13.8 which shows the pressure distribution on the model obtained from a RANS calculation. Based on these results, the user can choose the best configuration for the support without having to manufacture various stings and check the effects of each of them during an expensive wind tunnel tests.

The strong coupling between wind tunnel and CFD simulation also helps in guiding the CFD calculations for closer match with the experimental conditions. For instance, the calculation could be carried out by taking into account the model deformation under the effect of aerodynamic loads. The deformations are measured by stereo-imaging technique, by tracking the displacement of markers (see Sect. 9.6) and are used to deform the mesh on the surface of the CFD model. Also, while accounting for the test section walls and support effects it is possible to approach the conditions of the free flight.

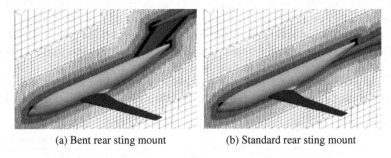

(a) Bent rear sting mount   (b) Standard rear sting mount

Fig. 13.7 Mesh for the RANS simulation of the effect of supports (© JAXA)

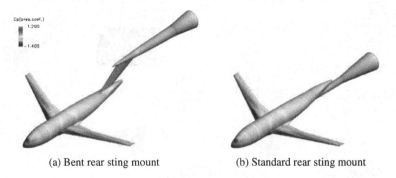

(a) Bent rear sting mount   (b) Standard rear sting mount

Fig. 13.8 Influence of the support of the model on the pressure distribution (© JAXA)

13.5 Reconstruction of Data

During experiments in many fields of applied science one is confronted by the fact
that the measured quantities give only a sparse representation of the event and it is
thus difficult to describe the physical mechanism in its entirety. This lack of resolution of the data can appear in various ways. For example, in fluid mechanics
measurements usually have an insufficient resolution either in time or space so contains partial information of the flow field. Moreover, the field of exploration might be
contaminated by measurement noise or data acquisition issues. The approach called
data reconstruction originates from the field of meteorology where the weather is
forecasted by analysing the trend of the atmospheric streams and is extrapolated in
time and space, based on sparse data resulting from various types of measurement in
stations distributed around the world. Today, data reconstruction has become quite
attractive in the field of fluid mechanics as it allows the analysis of a global flow field
through limited strategic local measurements.

Data assimilation consists of a calculation-experiment coupling having for objective to fulfil the gaps of the experiment, in particular the dispersed character of
the points of measurement and their insufficient density. Digital simulations can
take form of a simple constraint (for example, to interpolate between scattered
three-dimensional vectors, under the constraint that the interpolated field respects
the incompressible Navier–Stokes equations), or of a more powerful tool of super-
resolution (to find the initial and boundary conditions of a numerical simulation best
approaching all the available fields of vectors during a given temporal horizon). In
addition, an important stream of research aims at the estimate of the pressure field
starting from the velocity measurements, the knowledge of the velocity components
making it possible to calculate their gradients and thus evaluating the pressure. The
method applies either in a direct way in the case of flows with low Mach number, or
by introducing additional assumptions, such as for example the isentropic character,
in the case of compressible flows.

This technique can be applied in various ways, for instance the whole flow state
(pressure, velocity, temperature, etc.) can be estimated at one given instant over the
whole spatial domain and the evolution in time can be predicted. Mathematically
this falls into the same category of inverse problem where further information is
reconstructed from a limited number of measurements. Several methods have been
developed for data reconstruction ranging from a simple interpolation to more sophisticated approaches which exploit the equations controlling the system and in fluid
mechanics the Navier–Stokes equations are employed either averaged in time or fully
unsteady. The formulation of this problem is based on the variation of parameters
approach, highly studied in meteorology.

For example, let us consider a flow which can be described by the Reynolds
Averaged Navier–Stokes equations (RANS model) formulated like classical Navier–
Stokes equations but with a model for the Reynolds stress term. This term is an
unknown and is selected as a control parameter. An algorithm of data reconstruction
can be derived by minimising the error between the mean values of that quantity

measured in the flow and the values resulting from the numerical solution of the Navier–Stokes equations. Thus, this method can be used to improve the quality of results from PIV measurements, by spatial resolution refinement, extension of the field explored beyond the field of acquisition and the reconstruction of quantities not captured from the measurements. In this context, it is also necessary to cite the Proper Orthogonal Decomposition (POD) method, based on an incomplete decomposition which was employed successfully to rebuild missing PIV images.

In an unsteady framework, data reconstruction can be also used to carry out a noise filtering, to improve space and time super-resolution of insufficiently sampled PIV measurements. Within this framework, either the complete unsteady Navier–Stokes equations are used, or simplified models less expensive in computing times. As for the previous example, the optimisation algorithm is obtained by building a set of differential equations which expresses adequacy between simulated flow and the whole set of available data (i.e., all images available of the time series over which the reconstruction applies), with the constraint of respecting the governing equations chosen for the model. The control variables are normally the boundary and initial conditions of the simulation.

Such approaches were successfully applied within various frameworks: for instance in poorly seeded for time-resolved PIV measurements, where the reliability and accuracy of the data relies on tracking a large sample of particles (see Sect. 11.6.5). In the case of over-seeded PIV measurements, the data reconstruction technique allows for noise filtering and spatio-temporal extrapolation with super-resolution. Figure 13.9 presents the iso-contours of the transverse velocity component in a jet plane at a given instant in time; a comparison of the scattered initial PIV measurement against the reconstructed result using direct numerical simulation of the Navier–Stokes equations shows the benefit of this technique. From Fig. 13.9a the missing data outside the field of view was reconstructed and the additional flow

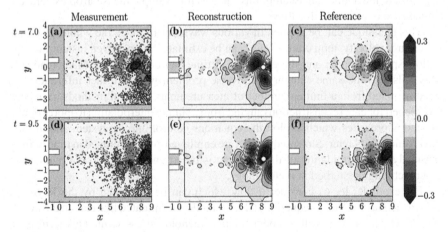

Fig. 13.9 Instantaneous iso-contours of the transverse velocity component of a plane jet (© Leclaire et al.)

features are illustrated in Fig. 13.9b. The assimilation makes it possible to determine the initial condition in the whole field as well as the velocity profile in the exhaust nozzle throughout the assimilated sequence. Note that in Fig. 13.9b the simulation allows an extrapolation and a significant filtering of the data which is important for analysis by searching for correlations and further calculations of quantities deduced from the velocity field.

Chapter 14
Prospects and Challenges for Aerodynamics

14.1 Role of the Wind Tunnel in Design and Optimisation

The wind tunnel is a major constituent of a specialised infrastructure for research in aeronautics and aerospace. Its contribution and importance over-match the experimental facilities required in many other sectors for research and development. In the absence of a complete tool set to design next generation vehicles, the design cycle can only be matured through joint CFD simulation and wind tunnel testing, where the latter is able to represent most of the physical phenomena that could be encountered during operation. This is a major source of disparity between CFD and wind tunnel results as CFD is still unable to reproduce or model the exact physics usually due to limitations in the choice of boundary conditions and realistic computational time and cost.

Aerial vehicles: The design process of modern aircraft is evolving with a closer integration of CFD simulation and wind tunnel testing through a more structured approach. The future design processes will focus on a more integrated aerodynamics and structure or aeroelasticity analysis with the objective of designing a lighter structure without compromising the safety. This tendency increases the need for perennial and adapted test facilities, measurement techniques of high quality for the development of highly accurate and robust simulation tools.

Therefore the main challenges in the aerospace sector, where experimentation and wind tunnels will play a crucial role, are:

- aeroacoustic measurements at high and low speeds in large wind tunnels in order to assess noise reduction techniques and ensure that the vehicle can pass certification procedures,
- integration of the engines in the wing and the fuselage to reduce drag,
- validation of the feasibility of the techniques for natural or hybrid laminar flow control in order to reduce the drag on wings while testing at the flight Reynolds number,

© Springer Nature Switzerland AG 2020
B. Chanetz et al., *Experimental Aerodynamics*,
Springer Tracts in Mechanical Engineering,
https://doi.org/10.1007/978-3-030-35562-3_14

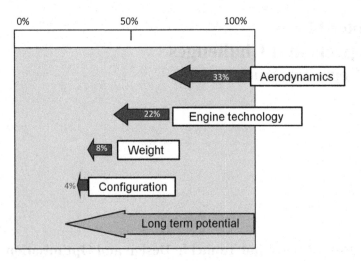

Fig. 14.1 Potential source for reduction in fuel consumption

- control of separation and vortices to reduce flow induced vibrations and once again noise,
- during the revival of supersonic and even hypersonic commercial aircraft projects,
- new requirements in terms of atmospheric re-entry resulting from the development of inhabited outer space.

However the major concerns remain the direct threat to the environment due to emissions from the aviation sector. Figure 14.1 shows a forecast by the European Union of the potential sources, through which the emission can be reduced. Aerodynamics represents the most promising and influential contribution with the potential reduction of 30%.

Terrestrial vehicles: for the terrestrial vehicles, the CO_2 emission standards are becoming increasingly strict, it will be essential to reconsider the use of wind tunnel, not only as a means of passing the safety requirements but also for performance optimisation through the reduction of drag, while providing other capabilities for assessing brake and under cap cooling, rainwater flow on windshield or in the vicinity of the engine, etc. The design of automobile results from compromise between several functionalities normally validated during wind tunnel tests. These would therefore govern their geometrical characteristics such as the size of wind tunnel test section or presence of rolling road, and if further information of the flow physics is required this will have an impact on the maximum operating speed and freestream turbulence intensity. This compromise is chosen through analysis of previous test results, numerical simulations, and based on engineering judgements developed from experience. But due to the large number of requirements and test conditions, it is difficult to converge towards the optimum trade-off. Some more recent wind tunnels, or those which have undergone significant development over time, allow for testing for a diverse range of requirements simultaneously. For example, the S2A wind tunnel at Saint-Cyr-l'École provides the opportunity of simultaneously studying the effect of drag and noise emitted in the A-pillar region, including the rear mirror strut (see Sect. 3.3.5). On the other hand, other conditions cannot be studied, for example braking, as the brakes would generate particles upon heating which would spoil the aeroacoustic treatment material of the wind tunnel. Other requirements such as rainy conditions cannot be tested either as the water could damage the balance used to measure aerodynamic forces.

The wind tunnel is only an approximation of the real conditions encountered on the road and therefore it must evolve to reproduce similar conditions such as high turbulence intensity, fluctuating wind, obstacles in the vicinity of the vehicle (lorries, trees, topology, etc.) These are challenges which will not fail to drive the passion of future aerodynamicist.

The issues to which aerodynamics are confronted are not limited and are spread over a large variety of disciplines. Current trends in the environmental impact due to transportation raise a lot of concerns which is driving the development of greener aviation and automotive industry. This should mobilise aerodynamicists in the years to come in three major fields of research, namely flow control, aeroacoustic and aerodynamic shapes optimisation, elaborated below.

14.2 Flow Control

Aerodynamics is essentially the study of the behaviour of the flow which would be later controlled or modified for a desired effect on the body moving in air or static but in moving air. The main objectives of controlling the flow are for the improvement of performance, increasing comfort and safety of the passengers, the reduction of unwanted unsteadiness such as vibrations, noise, vortices, splashes and above all the reduction of fuel consumption. The ability to control the flow relies on understanding the behaviour of the fluid at a very fundamental level. This involves advanced numerical calculations and fine experimental analyses based on the most sophisticated measurement techniques.

In aerodynamics the preferred phenomenon to be controlled are: laminar to turbulent transition, separation, turbulent flows, the interactions between shock wave and boundary layer, flow induced vibrations and noise.

For an aircraft in cruise the friction drag on the surface represents almost half of the total drag, therefore its reduction presents an opportunity for larger energy saving. Skin friction drag reduction can be achieved by maintaining the boundary layer in a laminar state by delaying the transition phenomenon. This is a complex flow instability phenomenon where theoretical and experimental research have been motivated to identify the factors influencing it, namely the ambient disturbances such as turbulence intensity and noise or the conditions on the surface such as pressure gradients, surface protuberances, vibration and heating. A laminar boundary layer can be sustained by tailoring a shape that promotes a favourable pressure gradient as well as improving the surface quality. Maintenance of a laminar state on the wing of an aircraft in operation can be compromised by the existence of defects on the surface, in particular at the interface of leading-edge slats or the anti-icing devices. A major issue is that the potential benefits of transition control technique cannot be easily assessed in classical wind tunnels due to higher levels of turbulence intensity and background noise (see Sects. 1.7.1 and 5.4).

The second aim of flow control is delaying or suppressing the separation of the boundary layer experiencing strong adverse pressure gradients, a rapid surface discontinuity (flows around corners and ramps) or in a shock wave. The presence of a separation almost always has detrimental consequences resulting in an increase in pressure drag, with the occurrence of large-scale fluctuations (buffeting on a wing), and a premature transition of laminar boundary layers. The drag of a terrestrial vehicle mainly constitutes of pressure drag, due to the lower pressure established in the separation region downstream of the vehicle (see Fig. 14.2). This drag contribution is particularly significant for square-back type vehicles.

Separation control is the subject of very active research where the aim is to implement means to suppress or delay separation by re-energising the boundary layer. This can be achieved through various processes, such as vehicle shape (morphing for example, see Sect. 3.1.10), solid or fluidic vortex generators (VGs), local suction or blowing, plasma actuators, but not limited to these. Similar processes are also used to control boundary layer and shock-wave interaction on transonic transport aircraft wing and engine intake or turbomachine. Figure 14.3a shows the surface flow pattern induced by VGs installed upstream of a shock wave in a transonic flow and Fig. 14.3b

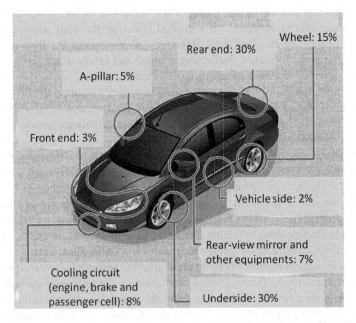

Fig. 14.2 Decomposition of an automobile aerodynamic drag (© PSA-Peugeot Citroën)

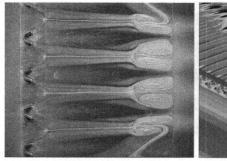 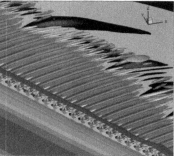

(a) Solid VGs in a transonic flow   (b) Fluidic VGs on an aircraft wing (CFD)

Fig. 14.3 Control of separation by vortex generators (VGs) (© ONERA)

the field produced by fluidic VGs placed close to the leading edge of a wing in order to control buffet (CFD results).

Last but not the least, the wing tip vortices of large transport aircraft are also the subject of major research because of the hazard they represent for an aircraft closely following or landing or taking off in the wake of these vortices. Thus, devices such as winglets or sharklets and wingtip fences placed at the extremity of the wing can control these vortices to a certain extent, but mainly helps in reducing the lift induced drag. Figure 14.4 shows a breakdown of the total drag of a transport aircraft and the potential techniques for drag reduction and the corresponding benefits.

Total drag (%)		Aerodynamics actions	Potential gain
100 — 80	parasitic interferences wave drag	○ shock control ○ integration	-3%
80 — 40	induced drag	○ shape optimisation ○ adaptive wing (morphing) ○ wing tip	-10%
40 — 0	friction drag	○ laminarity control ○ turbulence control ○ separation control	-20%
		Total	-33%

Fig. 14.4 Potential of drag reduction of a transport aircraft

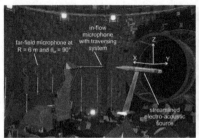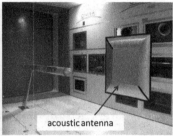

Fig. 14.5 Test of de-reverberation in CEPRA19 (left) and in non-anechoic closed test section in F1 wind tunnel (© ONERA)

14.3   Developments in Aeroacoustic Measurements

There is an increasing demand for acoustic measurements in wind tunnels with closed test sections where the main interest being the ability of undertaking simultaneous aerodynamic and acoustic measurements. This requirement is also justified by the need for measurements made in a closed field, free from the effect of the turbulence in the shear layer of an open test section. This leads to the microphone measurements being performed at the test section wall which this therefore made from high mechanical strength materials such as Kevlar or metal reinforced sheets.

Another issue with acoustic measurements in a closed test section arises from the effects of reverberation or wave reflections from walls which are not acoustically treated, which is the case in general. These contaminations can be isolated by employing advanced signal processing techniques. The first method is based on beam-forming using a network of microphones and allows the separation of the acoustic sources and hence identifying the real source from the reflected sources. The major constraint of this method is the significant number of microphones required to ensure a good spatial resolution for more accurate directivity measurements. Since the measurement points are fixed by the microphone array the resolution and field of measurement will be governed by the inter-microphone spacing and the distance from the source. Figure 14.5 illustrates the type of tests that should be carried out to determine the range of validity of this method, where the propagation of an electrically driven acoustic source is characterised first in an anechoic wind tunnel then in a wind tunnel with closed test section and rigid wall.

14.4   Search for Novel Aircraft Architectures

The objectives here are to reduce fuel consumption for the reduction of greenhouse gas emission and to decrease noise pollution. This is an all the more ambitious

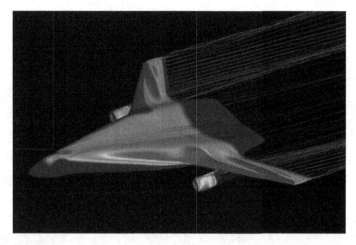

Fig. 14.6 Configuration of flying wing. project CICAV of ONERA (© ONERA)

challenge as the air traffic is expected to double in the next fifteen years. New break-through aircraft configurations with significant differences with the current ones, dating more than 50 years, are thus under investigation.

During a conceptual study at ONERA (the CICAV project), the flying wing config-uration was forecasted at succeeding the A350 and A320 NEO airplanes by the 2030–2050 horizon. As shown in Fig. 14.6, the flying wing consists of a blended wing-fuselage-engine configuration with the aim of saving the weight by approximately 10% and increase aerodynamic efficiency (lift to drag ratio) by 15%.

The civil aircraft project NOVA, led by ONERA, targets a flight range not currently addressed by medium-haul transport aircraft such as the A321-200 and the shorter-haul carriers like the B737-900ER. The range targeted is 5500 km, for 180 passengers (with seven passengers in front line with two aisles) and a cruising speed of 900 km/h (Mach 0.82). This concept presents a breakthrough in architecture due to its major differences with conventional configurations, constituting here of a wide composite lifting fuselage, with an oval section and a large aspect ratio wing with winglets oriented downwards to reduce the induced drag (see Fig. 14.7).

The characteristic of this architecture lies in the embedded large engines, towards the rear of the fuselage and capable of ingesting the boundary layer which develops upstream. The boundary-layer ingestion capability helps in reducing drag and hence reduction in greenhouse gas emissions. Propulsion efficiency increases with bypass ratio; however this introduces further compromise for embedded engines especially for boundary-layer ingestion types as the volume or cross-section tends to be a lot lower towards the trailing of fuselage due to the need of fairing for lower drag.

The potential benefits from electric propulsion is being assessed in projects such as AMPERE by ONERA-CEA, Vahana of Airbus, the Lilium Jet and Ehang 184, to mention a few. These vehicles are for intra or inter urban mobility, flying at a maximum altitude of 3000 m, at speeds of ranging around 250 km/h to cover a range

Fig. 14.7 Configuration with large lifting fuselage. NOVA project of ONERA (© ONERA)

Fig. 14.8 Project AMPERE (ONERA-CEA) of electric regional plane (© ONERA)

of 500 km which is significantly less than most of the current civil aircraft (see Fig. 14.8). To some extent, they could be considered as an air taxi, as they will be expected to take off and land in urban area and while carrying four to six passengers. They are expected to be unmanned vehicles for higher safety of operations with the possibility of vertical take-off and landing. The architectures of the proposed concepts vary drastically from conventional designs through implementation of several key technologies such as:

– distributed propulsion with electrically driven propellers,
– innovative configuration adapted to the distributed propulsion,
– hybrid energy sources,
– an integrated control including energy management and the possibilities of thrust vectoring.

14.5 Supersonic and Hypersonic Flights

Even after nearly 50 years since the first flight of Concorde and a little more than 20 years after its withdrawal from service, supersonic transport remains a major challenge for the civil aircraft industry. In the days of the Concorde, the technological achievements were through the design of an aerodynamic configuration suitable for supersonic cruise, capable of taking-off and landing on the existing runways and to develop materials resistant to the high stagnation temperatures at Mach 2 cruise. The design of the engine with optimum specific fuel consumption and its reliability over a range of more than 6000 km was also a true challenge. Since then, considerable progress has been made in materials science both for the body, through the use of titanium alloy, as well as for dealing with the high temperatures in the engines through the use of turbine blades manufactured through single-crystal. The power of CFD and the maintenance in service of large supersonic wind tunnels such as S2MA at ONERA should allow for the optimisation of a better aerodynamic configuration. Most influential aircraft manufacturers have a supersonic transport concept aircraft, but it should be noted that these projects are not only successors of Concorde but targets a different market which is corporate jets with maximum capacity of less than 20 passengers and it seems that this market is gaining a lot of interest. The targeted cruise Mach numbers are in the range of 1.4 to 2 for the best trade-off between flight duration and range.

With regards to aerodynamic performance, it is important to characterise the state of the flow on the wing which could be laminar at cruise conditions. The other challenges posed, are the direct environmental impact such as noise at take-off, propagation of the sonic boom to the ground and emissions of CO_2 and NO_x. There are strict regulations already in place and the increasing need to cut down emission could hinder the progress in the development of a new generation of supersonic aircraft. But, this also drives further research into more specific field such as the sonic boom phenomenon which prohibits supersonic flight over inhabited land. The Defence Advanced Research Projects Agency (DARPA) in the United States is trying to address this issue by studying the propagation of the sonic boom from the plane to the ground and aerodynamic configurations which would minimise its intensity (see Fig. 14.9). This work should also contribute to specifying a regulation on the acceptable level of sonic-boom intensity and open up the way for commercial supersonic transport. The extrapolation of the wind tunnel tests to characterise and evaluate the propagation of the real sonic boom in flight and its impact on ground will be then a new challenge.

Beyond the supersonic flight, there exist also concepts of hypersonic aircraft with trajectories inspired by intercontinental ballistic missiles and which could connect the antipodes in less than two hours. Concurrently to these very futuristic projects, one sees emerging the concept of "space tourism" which could represent a significant increase of the atmospheric re-entries from other inhabited planets (see Fig. 14.10). Such developments will require an upgrade and new developments of the hypersonic wind tunnels.

Fig. 14.9 Project of "low boom" supersonic aircraft of the DARPA

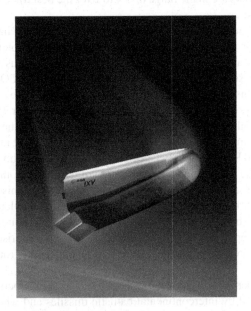

Fig. 14.10 Project IXV (Intermediate experimental Vehicle) of atmospheric re-entry shuttle (European demonstrator ESA/Alenia Space) (© Dassault Aviation)

14.6 Prospects for the Aerodynamic Design

As mentioned at the beginning of this book, the place of CFD in the aerodynamic design has become increasingly important. This is due to the exponential progress in computing powers and the techniques of numerical analysis allowing the resolution of increasingly sophisticated theoretical models. In parallel, wind tunnels, where real operating conditions can be approached, are being developed and the existing facilities are being equipped with more sophisticated means of positioning the models, measurements and analysis of the flow which has considerably increase their productivity. Thus, for several decades now the aerodynamic design of a vehicle (automobile or aircraft) has consisted of overlapping cycles.

CFD allows for computation of the flow field around the configuration of a vehicle in a few minutes and it is often more time consuming to define the geometry and generate the mesh than the actual computation. An aerodynamic design office can thus compute several configurations per day. There are even techniques for automatic shape optimisation which can reproduce the best shape according to the criterion defined by the user in few hours. This makes it possible to evaluate a very large number of variables within the space of possible configurations and a multi-disciplinary optimisation can be performed based on other constraints imposed by aeroelasticity or flight mechanics, for instance.

However, CFD is hindered by their lack of representativeness of complex flow phenomena. Competition in the aviation sector is driving more unconventional architectures that cannot be handled by the current tool set developed and validated from semi-empirical database; therefore this is proving to be a real issue right from the initial design phase. The wind tunnel tests are the means of studying a flow mechanism in a real environment, even if the conditions are not completely identical to the open, infinite conditions. The major disadvantages of the wind tunnel test for design and optimisation are the test durations, the costs of both model manufacturing and operation, including instrumentations. The duration of a campaign, including both model manufacturing and wind tunnel tests is typically of several months. If together with a baseline model another set of alternative configurations are manufactured, the test duration for the additional configurations can be reduced, since the test protocol has already been established. These test cases can be very useful, especially at conditions close to the edge of the flight envelope, where complex flow phenomena are encountered for which the representativeness of CFD is still questionable.

Wind tunnels can also be distinguished by their operational cost. The large wind tunnels in which the conditions closer to flight can be reproduced have significant operational costs. The models adapted for these tunnels are also very expensive. For these reasons, aircraft manufacturers often use smaller wind tunnels, less representative but cheaper, for the development tests. These wind tunnels are complementary to CFD since they provide test conditions for which CFD is least validated.

The more flight representative, large wind tunnels are then used to characterise more accurately the aerodynamic performance at the later stage of design or to explore very innovative technologies which will be implemented during optimisation and operated at cruise conditions, for instance laminar flow wings.

The rapid progress in CFD is quite often misinterpreted as a way forward for rendering wind tunnels obsolete. Indeed, if the processing power of computers continue to progress at the same rate as that over the last several decades, by a factor greater than 100 every 10 years, it is probable that the aerodynamicists starting their career will be able, to conduct a full Navier-Stokes simulation (DNS) on a complete aircraft configuration in a few hours. However, due to the stringent safety requirements in aviation and the lessons still to be learnt from the development of unconventional configurations, where different boundary conditions might emerge, it will be very risky to run a design campaign entirely on simulations. The coupled numerical modelling and experimentation is a complementary approach which increases the confidence level in the result. The wind tunnels are thus not intended to disappear but in fact to evolve.

In the short run, the industrial wind tunnel tests will be more focussed on productivity, by improving the capability to obtain an increasingly large amount of data and a variety of measurements for the same duration of test. Therefore, it implies the following:

- Integration of more sensors in the models and simultaneous optical flow diagnostic techniques on the model surface, using time-resolved PSP, TSP, and other laser-optical techniques such as LDV and PIV to mention a few.
- Motorisation of certain movable parts of the models, such as control surfaces, to limit the change-over of configurations, which result in down time of the wind tunnel; this will require further development in the control systems.

In the medium term, the innovative aircraft architectures will certainly include more integrated propulsion systems so the traditional split between drag and thrust will not be relevant. It will be then necessary to represent the flow through the propulsion systems in wind tunnel tests; this is already achieved by with TPS (see Sect. 4.3.2), but will become essential for almost all wind tunnel testing of full configurations. The models will be increasingly complex, therefore more expensive. Progress in manufacturing techniques, additive layer manufacturing in particular, should allow important advances in this field to allow for further miniaturisation of parts. Reduction of the production cycles of the models is also a challenge for productivity.

The rapid growth in aviation will probably lead to an increase in the number of airports and possibly night flights in the vicinity of urban areas. Due to the already stringent noise regulations imposed by the certification body further research in aeroacoustics will be favoured for the identification of the aerodynamic noise sources and to control it to acceptable certification levels. This will necessitate further progress in the means and techniques of acoustic signal capturing and treatment that could be implemented in the main wind tunnels facilities.

A generalisation of the techniques of real time adaptation of the test section walls (shape and/or permeability) to recreate unconfined conditions in flight would be particularly useful for transonic tests and could allow the testing of larger models (see Sect. 4.2.2).

Finally, the ultimate goal for aerodynamicists would be to remove the support of the models in the wind tunnel. Can a robust magnetic levitation/suspension (in Sect. 2.5.2) or similar contactless system be implemented, with sufficient power to retain the model at a desired equilibrium position at the centre of the test section? Such a concept will bring wind tunnel conditions closer to those at flight, and allow more representative transient or oscillatory behaviour studies, whereby complementing the coupled experimental and numerical approach to design more tightly.

Numerical techniques must also penetrate the wind tunnels physical environment. As presented in Chap. 13, these two means are complementary and essential to optimise the vehicle. However, they still suffer from lack of harmonisation. Why not design a wind tunnel with the means to carry out live computations within its control room? If the realisation of a CAD definition is a task incompatible with the duration of a test, on the other hand modification of a CAD definition, as well as the return of a computation, could be as fast as approximately one hour. As mentioned in Chap. 13 this capability of rapid numerical simulation and post processing could be very beneficial, in controlling and guiding the experimental campaign to increase productivity and accuracy. Certain steps must still be executed and be performed simultaneously trough automation, for instance rapid shape and deformation identification, and integration of the deformed shape within the reference CAD geometry of the model being tested or undergoing the computation. This synergy which is already in progress still suffers from limitations which are more academic than technological.

14.7 Aerodynamics and Teaching

The advent of the Internet has modified the diffusion of knowledge and pedagogy, not to forget increasing accessibility to black-box tools due to the rapid development in numerical methods is reshaping the engineering curriculum. It is not a question here of specifying in detail this transformation of the educational practice, but of treating an original way of sensitising high school and college students about aerodynamics: this is the intention of the EOLIA project supported by 3AF. The EOLIA wind tunnel was designed by the initiative of the 3AF Poitou Regional Group which built the first prototype in 2005 (see Fig. 14.11). In 2017, 23 schools in France and abroad had their own EOLIA wind tunnel. For the success of the EOLIA project the following specifications were at the heart of its design:

– Modular; so as to facilitate transportation of the wind tunnel,
– cost of materials and ease of machining,
– safety, so that it could be operated by young students with minimal supervision,
– a flow quality suitable for experiments of academic nature but which could be easily implemented.

The design of the wind tunnel is based on the three major components of the wind tunnel built by Gustave Eiffel in 1912: a collector, a test section and a diffuser equipped with a fan. This principle was retained so as to preserve the modular aspect

Fig. 14.11 Prototype of the teaching wind tunnel EOLIA

and for ease of assembly. The collector is an assembly of four plates folded along a frame which are used as stiffeners. The test section, with a square section of 30 cm and 75 cm long, is made of four detachable panels, for the ease of instrumentation and a speed of 0 to 25 m/s can be achieved in the test section. The wind tunnel has an overall length of 3.5 m.

The safety features put in place are: protection grid at the fan exit, manual speed control with at an automatic safe running velocity limit which cannot be exceeded, isolated electrical circuit up to the required standards and emergency stop button. Basic aerodynamic force measurements and visualisations by light sheet can be carried out in the test section.

The EOLIA wind tunnel is a pedagogical tool for colleges and high school students and for presentations to the general public, 3AF is committed to provide all the necessary drawing, advices and documentations, after-sales service and to contribute to an overall dynamics between the schools concerned. By this approach, 3AF intervenes in a constructive way in the world of teaching and assumes one of its functions as a scientific society of reference in the field of aeronautics and space.

Bibliography

Adrian RR, Durão DFG, Durst F, Heitor MV, Maeda M, Whitelaw JH (eds) (1992) Laser techniques and applications in fluid mechanics. In: Proceedings of the 6th international symposium, Springer-Verlag, Portugal, 20–23 July 1992

Boisson HC. et Crausse P (2014) De l'aérodynamique à l'hydraulique, un siècle d'études sur modèles réduits, éditions Cépaduès, ISBN 978.2.36493.093.3, réf. 1093

Chanetz B, et Coët MC (2004) Souffleries aérodynamiques. Encyclopedia Universalis, Cd-rom 10

Chometon F, Gilliéron P (1996) A survey of improved techniques for analysis of three-dimensional separated flows in automotive aerodynamics. SAE Congress, Detroit, Michigan, USA, 26–29 Feb 1996

Cousteix J (1989) Aérodynamique couche limite laminaire. Éditions Cépaduès. ISBN 2854282086, Réf. 100, 1989 ref. 200, 1989

Cousteix J (1989) Aérodynamique turbulence et couche limite. Éditions Cépaduès. ISBN 2854282108, Réf. 200, 1989

Délery J (2008) Handbook of compressible aerodynamics. Wiley

Délery J, Chanetz B (2000) Experimental aspects of code verification/validation: application to internal aerodynamics. VKI lectures series on verification and validation of computational fluid dynamics, 5–8 June 2000

Gilliéron P (2012) Aérodynamique instationnaire. Comprendre la méthode des caractéristiques. Éditions Cépaduès. ISBN 978.2.36493.010.0, Réf. 1010, 2012

Gilliéron P, et Kourta A (2011) Aérodynamique automobile pour l'environnement, le design et la sécurité. Editions Cépaduès. ISBN 978.2.36493.091.9, réf. 1091, 2011–2014

Hoerner SF (1993) Fluid-dynamic drag. Publisher Hoerner Fluid dynamics

Moisy F (2014) Méthodes expérimentales en mécanique des fluides. Master 1 de Physique Appliquée et Mécanique, Université Paris sud, 2014–2015

Pierre M (1995) Développement du centre d'essais de l'ONERA à Modane-Avrieux. Edition ONERA

Rebuffet P (1958) Aérodynamique expérimentale, vol 1, et 2. Dunod, Paris, 1958–1969

Tropea C, Yarin AL, Foss JF (eds) (2007) Handbook of experimental fluid mechanics. Springer-Verlag

© Springer Nature Switzerland AG 2020
B. Chanetz et al., *Experimental Aerodynamics*,
Springer Tracts in Mechanical Engineering,
https://doi.org/10.1007/978-3-030-35562-3

Chapter 1: The Experimental Approach in Aerodynamic Design

Albisser M (2015) Identification of aerodynamic coefficients from free flight data. Thèse de Doctorat de l'Université de Lorraine, 10 juillet 2015

Berner C, Dobre S, Albisser M (2015) Recent developments at the ISL open range test site and related measurement techniques. In: 66th aeroballistic range association meeting, San Antonio/TX, US, 4–9 Oct 2015

Chanetz B, Delery J, et Veuillot JP (2007) Article aérodynamique. Encyclopedia Universalis, édition 2007

Chanetz B (2015) Les souffleries. Article PANORAMA, la Science au Présent 2015, pp 180–193

Duncan GT, Crawford BK, Saric WS (2013) Flight experiments on the effects of step and gap excrescences on swept-wing transition. In: 48th 3AF international conference on applied aerodynamics, Saint-Louis, France, 25–27 Mar 2013

Giraud M (1982) Rôle du tunnel de tir balistique. Janvier, Revue de la Défense Nationale, pp 117–130

Chapter 2: Wind Tunnels and Other Aerodynamic Test Facilities

Chanetz B, Peter M (2013) Gustave Eiffel, a pioneer of aerodynamics. Int J Eng Syst Modell Simul 5(1/2):3

Theodule ML, Mannoni L, et Chanetz B (2004) Marey, précurseur oublié des souffleries. La Recherche n 380, pp 67–71, Nov 2004

Chapter 3: Subsonic Wind Tunnels

Chanetz B, Coët MC, et TENSI J (2010) La Grande soufflerie de Modane. PEGASE, la revue du Musée de l'Air, n 137, juin 2010

Eiffel G (1914) Nouvelles recherches sur la résistance de l'air et l'aviation faites au laboratoire d'Auteuil. In: Dunot H, et Pinat E (éditeurs)

Wind Tunnels of the PRISME Laboratory

Debien A, Von Krbek KAFF, Mazellier N, Duriez T, Cordier L, Noack BR, ABEL MW, Kourta A (2016) Closed-loop separation control over a sharp edge ramp using genetic programming. Exp Fluids 57:40

Kourta A, Thacker A, Joussot R (2015) Analysis and characterization of ramp flow separation. Exp Fluids 56:104

Muller Y, Aubrun S, Masson C (2015) Determination of real time predictors of the wind turbine wake meandering. Exp Fluids 56:53

Volpe R, Devinant P, Kourta A (2015) Experimental characterization of the unsteady natural wake of the full scale square-back Ahmed body: flow bi-stability and spectral analysis. Exp fluids 56:99

Wind Tunnels of the ONERA Lille Centre

Atinault O, Carrier G, Grenon R, Verbeke C, Viscat P (2013) Numerical and experimental aerodynamic investigations of boundary layer ingestion for improving propulsion efficiency of future air transport. In: 31st AIAA applied aerodynamics conference, San Diego, 24–27 June 2013

Mouton S, Rantet E, Gouverneur G, Verbeke C (2012) Combined wind tunnel tests and flow simulations for light aircraft performance prediction. In: 3AF 47th symposium of applied aerodynamics, Paris, 26–28 Mar 2012

Pape AL, Lienard C, Verbeke C, Pruvost M, De Coninck JL (2003) Helicopter fuselage drag reduction using active flow control: a comprehensive experimental investigation. J Am Helicopter Soc 60(3):1

Verbeke C, Mialon B (2014) Naval aerodynamics: State of the art measurement and computational techniques. In: 3AF 49th international symposium of applied aerodynamics, Lille, France, 24–26 Mar 2014

Verbeke C, Eglinger E, Atinault O, Grenon R, Carrier G, Mialon B, Ternoy F (2014) Experimental investigation of the boundary layer ingestion: specificity and challenges. In: 3AF 49th international symposium of applied aerodynamics, Lille, France, 24–26 Mar 2014

World-Directory (1990) World directory of aerospace vehicle research and development. World-Directory, Washington, DC

Wind Tunnels of the Institut de Mécanique des Fluides de Toulouse

Deri E, Braza M, Cazin S, Cid E, Degouet C, Michaelis D (2014) Investigation of the three-dimensional turbulent near-wake structure past a flat plate by tomographic PIV at high Reynolds number. J Fluids Struct 47:21–30

Mockett C, Perrin R, Reimann T, Braza M, Thiele F (2010) Analysis of detached-eddy simulation for the flow around a circular cylinder with reference to PIV data. J Flow, Turb Combust 85(2):167–180

Perrin R, Braza M, Cid E, Cazin S, Chassaing P, Mockett C, Reimann T, Thiele F (2008) Coherent and turbulent process analysis in the flow past a circular cylinder at high Reynolds number. J Fluids Struct 24(8):1313–1325

Scheller J, Chinaud M, Rouchon JF, Duhayon E, Cazin S, Marchal M, Braza M (2015) Trailing-edge dynamics of a morphing NACA0012 aileron at high Reynolds number by time-resolved PIV. J Fluids Struct 55:42–51

The CIRA Icing Wind Tunnel

Bellucci M, Esposito BM, Marrazzo M, Esposito B, Ferrigno F (2007) Calibration of the CIRA IWT in low speed configuration. In: 45th AIAA aerospace sciences meeting and exhibit, Reno, Nevada, AIAA-2007-1092, 8–11 Jan 2007

Esposito BM, Ragni A, Ferrigno F, Vecchione L (2003) Update on the icing cloud calibration of the CIRA Icing Wind Tunnel. SAE 2003 Transactions, J Aerosp, vol 112, Sec 1, p 47, Apr 2003

Oleskiw MM, Esposito BM, de Gregorio F (1996) The effect of altitude on icing tunnel airfoil icing simulation. In: Proceedings of the FAA international conference on aircraft inflight icing, DOT/FAA/AR-96/81, II, pp 511–520, Aug 1996

Ragni A, Esposito BM, Marrazzo M, Bellucci M, Vecchione L (2005) Calibration of the CIRA IWT in the high speed configuration. In: 43rd AIAA aerospace sciences meeting and exhibit, Reno, Nevada, AIAA 2005-471, 10–13 Jan 2005

Wind Tunnels of the PPRIME Institute at Poitiers

Cavalieri AVG, Rodriguez D, Jordan P, Colonius T, Gervais Y (2013) Wave packets in the velocity field of turbulent jets. J Fluid Mech 730:559–592

Jordan P, Gervais Y (2008) Subsonic jet aeroacoustics: associating experiment, modelling and simulation. Invited review. Exp Fluids, vol 44

Koenig M, Sasaki K, Cavalieri AVG, Jordan P, Gervais Y (2016) Jet-noise control by fluidic injection from a rotating plug: linear and non-linear sound-source mechanisms. J Fluid Mech, vol 788

Koenig M, Cavalieri AVG, Jordan P, Delville J, Gervais Y, Papamoschou D (2013) Farfield filtering and source imaging of subsonic jet noise. J Sound Vib 332(18):4067–4088

Laurendeau E, Jordan P, Bonnet JP, Delville J, Parnaudeau P, Lamballais E (2008) Subsonic jet noise reduction by fluidic control: the interaction region and the global effect. Phys Fluids 20:101519

Laurendeau E, Jordan P, Delville J, Bonnet JP (2008) Source identification by nearfield-farfield pressure correlations in subsonic jets. Int J Aeroacoust 7(1):41

Lazure H, Moriniere V, Laumonier J, et Philippon L (2016) Sifflement aérodynamique d'un rétroviseur. In: 13ème Congrès Francais d'Acoustique, Le Mans, pp. 1891–1897 [N°000383], 11–15 avril 2016

Marchiano R, Druault P, Leiba R, Marchal J, Ollivier F, Valeau V, et Vanwynsberghe C (2016) Localisation de sources aéroacoustiques par une méthode de retournement temporel tridimensionnelle. In: 13ème Congrès Français d'Acoustique, Le Mans, [N°000353]:1899–1905, 11–15 avril 2016

Maury R, Koenig M, Cattafesta L, Jordan P, Delville J (2012) Extremum-seeking control of jet noise. Int J Aeroacoust 11(3):459

Semeraro O, Jaunet V, Lesshafft L, Jordan P (2016) Modelling of coherent structures in a turbulent jet as global linear instability wave packets: theory and experiment. Int J Heat Fluid Flow, vol 62

Tissot G, Zhang M, Lajus FC, Cavalieri AVG, Jordan P (2017) Sensitivity of wave packets in jets to nonlinear effects: the role of the critical layer. J Fluid Mech, vol 811

Van Herpe F, Totaro N, Lafont T, Lazure H, et Laumonier J (2016) Prédiction de bruit d'origine aérodynamique en moyennes fréquences par la méthode SmEdA. In: 13ème Congrès Français d'Acoustique, Le Mans, pp. 1453–1459 [N°000374], 11–15 avril 2016

Wind Tunnels of IAT at Saint-Cyr-l'École

Hlevca D, Degeratu M, Grasso F (2014) Experimental and numerical modelling of atmospheric boundary layer development in a short wind tunnel. In: 3AF 49th international symposium on applied aerodynamics, Lille, 24–26 Mar 2014

Hlevca D, Gillieron P, Grasso F (2015) Active control applied to the flow past a backward facing ramp: S-PIV measurements and POD analysis. In: 3AF 50th international symposium on applied aerodynamics, Toulouse, 29–30 Mar 2015

Joseph P, Amandolèse X, Edouard C, Aider JL (2013) Flow control using MEMS pulsed micro-jets on the Ahmed body. Exp Fluids 54(1):1–12

Joseph P, Bortolus D, Grasso F (2014) Flow control on a 3D backward facing ramp by pulsed jets. Comptes Rendus Mécanique 342(6):376–381

Noger C, van Grevenynghe E (2011) On the transient aerodynamic forces induced on heavy and light vehicles in overtaking processes. Int J Aerodyn 1(3/4):373–383

The Hydrodynamic Channels of the Haut-de-Frances Polytechnic University

Hanratty TJ, Campbell JA (1983) Measurement of wall shear stress. In: R.J. Goldstein (ed) Fluid mechanics measurements. Hemisphere Publishing Co.
Fourrié G, Boussemart D, Keirsbulck L, et Labraga L (2011) Mesure du frottement pariétal instationnaire autour d'un corps épais 3D par méthode électrochimique. Congrès Français de Mécanique

Chapter 4: Transonic Wind Tunnels

Wind Tunnel S3Ch of the ONERA Meudon Centre

Brion V, Dandois J, Abart JC, Paillart P (2017) Experimental analysis of the shock dynamics on a transonic laminar airfoil. In: 7th European conference for aeronautics and space sciences, Milan, Italy
Bur R, Brion V, Molton P (2014) An overview of recent experimental studies conducted in ONERA S3Ch transonic wind tunnel. In: 29th congress of the international council of the aeronautical sciences (ICAS), St. Petersburg (Russia), 7–12 Sept 2014
Le Sant Y, Bouvier F (1992) A new adaptative test section at ONERA Chalais-Meudon. European forum on wind tunnels and wind tunnel test techniques, Southampton University (UK), (ONERA TP n° 1992–117), 14–17 Sept 1992
Molton P, Dandois J, Lepage A, Brunet V, Bur R (2013) Control of buffet phenomenon on a transonic swept wing. AIAA J 51(4):761–772. https://doi.org/10.2514/1.J051000

Chapter 5: Supersonic Wind Tunnels

Wind Tunnel S8Ch of the ONERA Meudon Centre

Bur R, Coponet D, Carpels Y (2009) Separation control by vortex generator devices in a transonic channel flow. Shock Waves J 19(6):521–530. https://doi.org/10.1007/s00193-009-0234-6
Merienne MC, Molton P, Bur R, Le Sant Y (2015) Pressure-sensitive paint application to an oscillating shock wave in a transonic flow. AIAA J 53(11):3208–3220. https://doi.org/10.2514/1.J053744
Molton P, Hue D, Bur R (2015) Drag induced by flat-plate imperfections in compressible turbulent flow regimes. J Aircraft 52(2):667–679. https://doi.org/10.2514/1.C032911
Sartor F, Mettot C, Bur R, Sipp D (2015) Unsteadiness in transonic shock-wave/boundary-layer interactions: experimental investigation and global stability analysis. J Fluid Mech 781:550–577. https://doi.org/10.1017/jfm.2015.510

Sartor F, Losfeld G, Bur R (2012) PIV study on a shock-induced separation in a transonic flow. Exp
    Fluids 53(3):815–827. https://doi.org/10.1007/s00348-012-1330-4

Wind Tunnel S8 of IUSTI at Marseille

Agostini A, Larchevêque L, Dupont P (2015) Mechanism of shock unsteadiness in separated
    shock/boundary-layer interactions. Phys Fluids 27(12):126103
Debiève JF, Dupont P (2009) Dependence between shock and separation bubble in a shock
    wave/boundary layer interaction. Shock Waves 19(6):499–506
Dupont P, Haddad C, Debiève JF (2006) Space and time organization in a shock induced boundary
    layer. J Fluid Mech 559:255–277
Jaunet V, Debiève JF, Dupont P (2014) Length scales and time scales of a heated shock-
    wave/boundary-layer interaction. AIAA J, pp 1–9, 30 Aug 2014
Schreyer AM, Larchevêque L, Dupont P (2015) Method for spectra estimation from high-speed
    experimental data. AIAA J, pp 1–12, 18 Nov 2015
Souverein LJ, Bakker PG, Dupont P (2013) A scaling analysis for turbulent shock wave boundary
    layer interactions. J Fluid Mech 714:505–535

The Quiet Tunnel of Purdue University

Durant A, André T, Edelman JB, Chynoweth BC, Schneider SP (2015) Mach 6 quiet tunnel laminar
    to turbulent investigation of a generic hypersonic forebody. In: International space planes and
    hypersonic systems and technologies conferences, 20th AIAA international space planes and
    hypersonic systems and technologies conference, Glasgow, Scotland, 6–9 July 2015
Schneider SP (2008) Development of hypersonic quiet tunnels. J Spacecraft Rockets, vol 45, no 4,
    July–Aug 2008
Schneider SP (2015) Developing mechanism-based methods for estimating hypersonic boundary-
    layer transition in flight: the role of quiet tunnels. Progr Aerospace Sci 72:17–29

Chapter 6: Hypersonic Wind Tunnels

Chanetz B, Chpoun A, Supersonic and hypersonic wind tunnels. In: Ben-Dor G, Igra O, Elperin T
    (eds) (2001) Handbook of shock waves, vol 1, Chap 4.5. Academic Press, London, San Diego
Chanetz B, Coët MC, Nicout D, Pot T, Broussaud P, François G, Masson A, Vennemann D (1998)
    New hypersonic experiment means developed at ONERA: the R5Ch and F4 facilities. In: Pro-
    ceedings AGARD conference on theoretical and experimental methods in hypersonic flows,
    514
Lago V, Chpoun A, Chanetz B (2012) Shock waves in hypersonic rarefied flows. In: Brun R (ed)
    High temperature phenomena, pp 271–298, Springer

ISL Wind Tunnels

Gnemmi P, Srulijes J, Seiler F, Sauerwein B, Bastide M, Rey C, Wey P, Martinez B, Albers H, Schlöffel G, Hruschka R, Gauthier T (2016) Shock tunnels at ISL. In: Igra O, Seiler F (eds) Experimental methods of shock wave research. Springer Verlag. ISBN 978-3-319-23745-9

Gnemmi P, Rey C (2015) Experimental investigations on a free-flying supersonic projectile model submitted to an electric discharge generating plasma. In: 30 international symposium on shock waves (ISSW30), Tel Aviv, Israel, 19–24 July 2015

Wey P, Bastide M, Martinez B, Srulijes J, Gnemmi P (2012) Determination of aerodynamic coefficients from shock tunnel free flight trajectories. In: 28th aerodynamic measurement technology, ground testing, and flight testing conference, New Orleans/LO, AIAA Paper 2012-3321, 25–28 June 2012

Blow Down Wind Tunnels of the ONERA Meudon Centre

Bur R, Chanetz B (2009) Experimental study on the PRE-X vehicle focusing on the transitional shock-wave/boundary-layer interactions. Aerospace Sci Technol 13(7):393–401, Oct–Nov 2009. https://doi.org/10.1016/j.ast.2009.09.002

Benay R, Chanetz B, Mangin B, Vandomme L, Perraud J (2006) Shock-wave/transitional boundary-layer interactions in hypersonic flow. AIAA J 44(6):1243

Wind Tunnels of CIRA at Capua

Purpura C, de Filippis F, Graps E, Trifoni E, Savino R (2007) The GHIBLI plasma wind tunnel: Description of the new CIRA—PWT facility. Acta Astronautica 61(1–6):331–340

Russo G, De Filippis E, Borrelli S, Marini M, Caristia S (2002) The SCIROCCO 70-MW plasma wind tunnel: a new hypersonic capability. In: Lu FK, et Dan EM (eds) Advanced hypersonic test facilities, AIAA, pp 313–351, ISBN 1-56347-541-3

Trifoni E, Del Vecchio A, Di Clemente M, De Simone V, Martucci A, Purpura C, Savino R, Cipullo A (2011) Design of a scientific experiment on EXPERT flap at CIRA SCIROCCO plasma wind tunnel. In: 7th European symposium on aerothermodynamics for space vehicles, Bruges, Belgium, 9–12 May 2011

Trifoni E, Purpura C, Martucci A, Graps E, Schettino A, Battista F, Passaro A, Baccarella D, cristofolini A, Neretti G (2011) MHD experiment at CIRA GHIBLI plasma wind tunnel. In: 7th European symposium on aerothermodynamics for space vehicles, Bruges, Belgium, 9–12 May 2011

The HEG Wind Tunnel of DLR at Göttingen

Hannemann K (2003) High enthalpy flows in the HEG shock tunnel: experiment and numerical rebuilding. In: AIAA paper 2003-0978, 41st AIAA aerospace sciences meeting and exhibit, Reno, Nevada, 6–9 Jan 2003

Hannemann K, Martinez Schramm J, Brück S, Longo JMA (2001) High enthalpy testing and CFD.
    Rebuilding of X-38 in HEG. In: 4th European symposium on aerothermodynamics for space
    vehicles, CIRA, Capua, Italy, 15–18 Oct 2001
Martinez Schramm J, Karl S, Hannemann K, Steelant J (2008) Ground testing of the HyShot II,
    scramjet configuration in HEG. In: AIAA paper 2008-2547, 15th AIAA international space planes
    and hypersonic systems and technologies conference, Dayton, 28 Apr–1 May 2008
Sagnier P, Vérant JL (1998) On the validation of high enthalpy wind tunnel simulations. Aerospace
    Sci Technol 7:425
Sagnier P, Ledy JP, Chanetz B (2001) ONERA wind tunnel facilities for re-entry vehicle applications.
    In: 3AF 37th symposium on applied aerodynamics, Arcachon, France
Stalker RJ (1967) A study of the free-piston shock tunnel. AIAA J 5(12):2160

Chapter 7: Flow Visualisation Techniques

Chometon F, et Gilliéron P (1994) Dépouillement assisté par ordinateur des visualisations pariétales
    en aérodynamique. *Compte-Rendu de l'Académie des Sciences,* Paris, tome 319, pp 1149–1156
Délery J (2013) Three-dimensional separated flow topology. Wiley
Seiler F (2010) Flow visualization at high atmospheric altitude conditions in a shock tube. In:
    ISFV14—14th international symposium on flow visualization, EXCO Daegu, Korea, 21–24 June
    2010
Settles GS (2001) Schlieren and shadowgraph techniques: visualizing phenomena in transparent
    media. Springer-Verlag, Berlin
Smeets G (1990) Interferometry, lecture series 1990-05 on measurement techniques for hypersonic
    flows. von Karman Institute for Fluid Dynamics, Rhode-St-Genèse, Belgium, 28 May–1 June
    1990
Weinstein LM (1991) An improved large-field focusing Schlieren system. AIAA Paper 91-0567

Chapter 8: Measurement of Aerodynamic Forces and Moment

Amant S (2002) Calcul et décomposition de la traînée aérodynamique des avions de transport à
    partir de calculs numériques et d'essais en soufflerie, Mémoire de Thèse, ENSAE
Destarac D (1995) Evaluation de la traînée à partir de calculs Euler. ONERA RTS n° 52/3423AY,
    décembre 1995
Onorato M, Costelli A, Garonne A (1984) Drag measurement through wake analysis. SAE, SP-569,
    international congress & exposition, Detroit, Michigan, 27 Feb–2 Mar, pp 85–93

Chapter 9: Characterisation of Flow Properties at the Surface

Acharaya M, Bornstein J, Escudier MP, Vorkuka V (1985) Development of a floating element for
    the measurement of surface shear stress. AIAA J 23(3):410–415
Adrian RJ, Westerweel J (2011) Particle image velocimetry (no. 30). Cambridge University Press

Champagnat F, Plyer A, le Besnerais G, Leclaire B, Davoust S, le Sant Y (2011) Fast and accurate PIV computation using highly parallel iterative correlation maximization. Exp Fluids 50(4):1169–1182

Bouchardy AM, Durand G, Gauffre G (1983) Processing of infrared thermal images for aerodynamics research. In: 1983 SPIE international technical conference, Genève, 18–22 Apr 1983

Carlomagno G, Luca L (1986) Heat transfer measurements by means of infraredthermograph. In: Fourth international symposium on flow visualization, Paris, 26–29 Aug 1986

Crites RC (1993) Pressure sensitive paint technique. VKI lecture series 1993-05 on measurement techniques, Apr 1993

Gaudet L, Gell TG (1989) Use of liquid crystals for qualitative and quantitative 2D studies of transition and skin friction. In: RAE, Technical Memorandum Aero 2159, June 1989

Le Sant Y, Fontaine J (1997) Application of infrared measurements in the ONERA's wind tunnels. In: Wind tunnels and wind tunnel test techniques. Cambridge, U.K., 14–16 Apr 1997

Le Sant Y, Merienne MC (2005) Surface pressure measurements by using pressure-sensitive paints. Aerospace Sci Technol 9(4):285–299

Liu T (2004) Pressure-and temperature-sensitive paints. Wiley & Sons, Ltd.

Mébarki Y, Peintures sensibles à la pression: application en soufflerie aérodynamique. Ph. D. Dissertation, University of Lille 1, Mar 1998

Mébarki Y, Mérienne MC (1998) PSP application on a supersonic aerospike nozzle. PSP Workshop, Seattle, USA, 6–8 Oct 1998

Mérienne MC, Bouvier F (1999) Vortical flow field investigation using a two-component pressure sensitive paint at low speed. ICIASF 99, Toulouse, France, 14–17 June 1999

Monson DJ (1883) A laser interferometer for measuring skin friction in three-dimensional flow. In: AIAA paper 83-0385, Jan 1883

Seto J, Hornung H (1993) Two-directional skin frictions measurement utilizing a compact internally-mounted thin-liquid skin friction meter. In: AIAA paper 93-0180, Jan 1993

Settles GS (1986) Recent skin friction techniques for compressible flows. In: AIAA paper 86-1099, May 1986

Chapter 10: Intrusive Measurement Techniques

Bestion D, Gaviglio J, Bonnet JP (1983) Comparison between constant current and constant temperature hot-wire anemometers in high speed flows. Rev Scientif Instr 54(11):1513–1524

Bruun HH (1995) Hot-wire anemometry. Principle and signal analysis. Oxford Science Publications, Oxford University Press

Gaillard R (1990) Development of a calibration bench for small anemoclinometer probes. In: Symposium on measuring techniques for transonic and supersonic flow in cascades and turbomachines, Rhodes-Saint-Genèse (Belgium), 17–19 Sept 1990

Lomas CG (1986) Fundamentals of hot wire anemometry. Cambridge University Press

Thermal Anemometry (2007) Tropea, Yarin, Foss (eds) Chapitre 5.2 dans Springer handbook of experimental fluid mechanics, Springer

Tutkun M, George WG, Delville J, Foucalt JM, Coudert S, Stanislas M (2008) Space-time correlations from a 143 how-wire rake in a high Reynolds number turbulent boundary layer. AIAA paper 2008-4239

Chapter 11: Non-intrusive Measurement Techniques

Boutier A (1999) Caractérisation de la turbulence par vélocimétrie laser. In: 3AF 35ème Colloque d'Aérodynamique Appliquée, Lille, France, 22–24 mars 1999

Boutier A, Micheli F (1996) Laser anemometry for aerodynamics flow characterization. La Recherche Aérospatiale, n 3, pp 217–226

Davoust S, Jacquin L, Leclaire B (2012) Dynamics of m = 0 and m = 1 modes of streamwise vortices in a turbulent axisymmetric mixing layer. J Fluid Mech 709:408–444

Délery J., Surget J, Lacharme JP (1977) Interférométrie holographique quantitative en écoulement transsonique bidimensionnel. La Recherche Aérospatiale, n° 1977-2, pp 89–101

Desse J-M (1990) Instantaneous density measurement in two-dimensional gas flow by high speed differential interferometry. Exp Fluids 9(1–2):85–99

Desse J-M (1998) Effect of time-varying wake flow characteristic behind flat plates. AIAA J 36(11):2036–2043

Goldstein R-J (1974) Measurement of fluid velocity by laser Doppler techniques. Appl Mech Rev 27:753–760

Lempereur C, Barricau P, Mathe JM, Mignosi A (1999) Doppler global velocimetry: accuracy test in a wind tunnel, ICIASF 99, Toulouse, France, 14–17 June 1999

Meyers JF (1994) Development of doppler global velocimetry for wind tunnel testing. In: 18th aerospace ground testing conference, Colorado Springs, CO, AIAA Paper 94-2582, June 1994

Raffel M, Willert C, Wereley S, Kompenhans J (2007) Particle image velocimetry: a practical guide. Springer

Riethmuller ML (1997) Vélocimétrie par images de particules ou PIV. Ecole d'été de l'Association Francophone de Vélocimétrie Laser, Saint-Pierre d'Oléron, 22–26 Sept 1997

Roehle I, Wilhert C, Shodl R (1998) Application of 3D doppler global velocimetry in turbomachines. In: 8th international symposium on flow visualisation

Samimy M, Wernet MP (2000) Review of planar multiple-component velocimetry in high speed flows. AIAA J 38(4):553

Scarano F (2013) Tomographic PIV: principles and practice. Measurement Sci Technol, vol 24

Surget J, Délery J, Lacharme JP (1977) Holographic interferometry applied to the metrology of gaseous flows. First European congress on optics applied to metrology, Strasbourg, France, 26–28 Oct 1977

Vest C-M (1979) Holographic interferometry. Wiley-Interscience, New-York

Yanta WJ (1973) Turbulence measurements with a laser Doppler velocimeter. NOLTR, vol 73–94

Chapter 12: Laser Spectroscopy and Electron Beam Excitation

Beck WH, Trinks O, Mohamed A (1999) Diode laser absorption measurements in high enthalpy flows: HEG free stream conditions and driver gas arrival. In: 22nd international symposium on shock waves, imperial college, London, U.K., 18–23 July 1999

Gorchakova N, Kuznetsov L, Rebrov A, Yarigin V (1985) Electron beam diagnostics of high temperature rarefied gas. In: 13th International symposium on rarefied gas dynamics, vol 2, Plenum Press (eds), pp 825–832

Gorchakova N, Chanetz B, Kuznetsov L, Pigache D, Pot T, Taran JP, Yarigin V (1999) Electron beam excited X-ray method for density measurements of rarefied gas flows near models. In: 21st International symposium on rarefied gas dynamics, Cépaduès (eds), vol 2, pp 617–624

Grisch F, Bouchardy P, Péalat M, Chanetz B, Pot T, Coët MC (1993) Rotational temperature and density measurements in a hypersonic flow by dual-line CARS. Appl Phys B 56:14–20

Gorchakova N, Kuznetsov L, Yarygin V, Chanetz B, Pot T, Bur R, Taran JP, Pigache D, Schulte D, Moss J (2002) Progress in shock wave/boundary layer interactions studies in rarefied hypersonic flows using electron beam excited X-ray detection. AIAA J

Gross KP, Mc Kenzie RL, Logan P (1987) Measurements of temperature, density, pressure, and their fluctuations in supersonic turbulence using laser-induced fluorescence. Exp Fluids 5:372–380

Hiller B, Hanson RK (1988) Simultaneous planar measurements of velocity and pressure fields in gas flows using laser-induced fluorescence. Appl Opt 27(1):33–48

Kuznetsov L, Rebrov A, Yarigin V (1973) Diagnostics of ionized gas by electron beam in X-ray spectrum range. In: 11th international conference on phenomena in ionized gases, Prague

Larigaldie S, Bize D, Mohamed AK, Ory M, Soutadé J, Taran JP (1998) Velocity measurement in high enthalpy, hypersonic flows using an electron beam assisted glow discharge. AIAA J 36(6):1061

Lefebvre M, Chanetz B, Pot T, Bouchardy P, Varghese P (1994) Measurement by coherent anti-sokes Raman Scattering in the R5Ch hypersonic wind tunnel. Aerosp Res (4):295–298

Mohamed AK, Pot T, Chanetz B (1995) Diagnostics by electron beam florescence in hypersonics. In: 16th international congress on instrumentation in aerospace facilities, Dayton, OH, 18–21 July 1995

Chapter 13: Computer-Aided Wind Tunnel and Analysis

Esteve MJ, Esteve N (2012) CFD/WTT synergy towards an enhanced A/C performance prediction at Airbus. In: 3AF 47th international conference on applied aerodynamics, Paris, 26–28 Mar 2012

Foures DPG, Dovetta N, Sipp D, Schmid PJ (2014) A data-assimilation method for Reynolds-averaged Navier-Stokes-driven mean flow reconstruction. J Fluid Mech 759:404–431

Hantrais-Gervois JL, Piat JF (2012) A methodology to derive wind tunnel wall corrections from RANS simulations. In: 5th symposium on integrating CFD and experiments in aerodynamics (integration 2012), JAXA Chofu Aerospace Center, Tokyo, Japan, 3–5 Oct 2012

Watanabe S, Kuchi-Ishi S, Murakami K, Hashimoto A, Kato H, Yamashita T, Yasue K, Imagawa K, Nakakita K (2012) Development status of a prototype system for EFD/CFD integration. In: 3AF 47th international conference on applied aerodynamics, Paris, 26–28 Mar 2012

Chapter 14: Prospects and Challenges for Aerodynamics

Arnal D (2008) Practical transition prediction methods: Subsonic and transonic flows. VKI lecture series advances in laminar-turbulent transition modelling

Marchiano R, Druault P, Leiba R, Marchal J, Ollivier F, Valeau V. et Vanwynsberghe C (2016) Localisation de sources aéroacoustiques par une méthode de retournement temporel tridimensionnelle. In: 13ème Congrès Français d'Acoustique, Le Mans, pp 1899–1905 [N°000353], 11–15 Avril 2016

Van Herpe F, Totaro N, Lafont T, Lazure H, et Laumonier J (2016) Prédiction de bruit d'origine aérodynamique en moyennes fréquences par la méthode SmEdA. In: 13ème Congrès Francais d'Acoustique, Le Mans, pp 1453–1459 [N°000374], 11–15 Avril 2016

Vermeersch O, Yoshida K, Ueda Y, Arnal D (2012) Transition prediction on a supersonic natural laminar flow wing: experiments and computations. In: 3AF 47th conference on applied aerodynamics, Paris, 26–28 Mar 2012

Printed in the United States
By Bookmasters